Automated Driving and Driver Assistance Systems

Automated vehicles are set to transform the world. Automated driving vehicles are here already and undergoing serious testing in several countries around the world. This book explains the technologies in language that is easy to understand and accessible to all readers. It covers the subject from several angles but in particular shows the links to existing ADAS technologies already in use in all modern vehicles. There is a lot of hype in the media at the moment about autonomous or driverless cars, and while some manufacturers expect to have vehicles available from 2020, they will not soon take over and it will be some time before they are commonplace. However, it is very important to be ready for the huge change of direction that automated driving will take. This is the first book of its type available and complements Tom Denton's other books.

Tom Denton is the leading UK automotive author with a teaching career spanning from lecturer to head of automotive engineering in a large college. His portfolio of automotive textbooks published since 1995 are bestsellers and led to his authoring of the Automotive Technician Training multimedia system that is in common use in the UK, USA and several other countries. He is currently working with ELTE (largest university in Hungary) on an autonomous driving project, and as a consultant to the Institute of the Motor Industry.

Automated Driving and Driver Assistance Systems

Tom Denton

Routledge
Taylor & Francis Group

LONDON AND NEW YORK

First published 2020
by Routledge
2 Park Square, Milton Park, Abingdon, Oxon OX14 4RN

and by Routledge
52 Vanderbilt Avenue, New York, NY 10017

Routledge is an imprint of the Taylor & Francis Group, an informa business

British Library Cataloguing-in-Publication Data
A catalogue record for this book is available from the British Library

Library of Congress Cataloging-in-Publication Data
A catalog record has been requested for this book

ISBN: 978-0-367-26560-1 (hbk)
ISBN: 978-0-367-26559-5 (pbk)
ISBN: 978-0-429-29385-6 (ebk)

Typeset in Arial
by Apex CoVantage, LLC

Contents

v

Preface

In this book you will find lots of useful and interesting information about automated driving vehicles (ADVs). It is the sixth in the series:

▶ *Automobile Mechanical and Electrical Systems*
▶ *Automobile Electrical and Electronic Systems*
▶ *Automobile Advanced Fault Diagnosis*
▶ *Electric and Hybrid Vehicles*
▶ *Alternative Fuel Vehicles*
▶ *Automated Driving and Driver Assistance Systems*

Ideally, you will have studied the electrical book, or have some experience, before reading this one. But don't worry if not, you will still learn a lot from it. Because these new vehicles are not in common use yet, I have concentrated on outlining their systems and the associated issues. More details on service and repair will be added in future editions.

Because ADVs are new, I have also looked at some of the social and human issues related to this exciting change. Case studies are included to set some of the technologies in a time context.

Comments, suggestions and feedback are always welcome at my website:

www.automotive-technology.org

On this site, you will also find lots of **free** online resources to help with your studies. Check out the final chapter for more information about the amazing resources to go with this and my other books. These resources work with the book, and are ideal for self-study or for teachers helping others to learn.

Good luck with your studies, and I hope you find automotive technology as interesting as I still do.

Tom Denton
BA FIMI MSAE MIRTE Cert Ed

Acknowledgements

Over the years many people have helped in the production of my books. I am therefore very grateful to the following companies who provided information and/or permission to reproduce photographs and/or diagrams:

AA
AC Delco
ACEA
Alpine Audio Systems
Audi
Autologic Data Systems
BMW UK
Bosch
Brembo brakes
C&K Components
Citroën UK
Clarion Car Audio
Continental
CuiCAR
Dana
Delphi Media
Eberspaecher
First Sensor AG
Fluke Instruments UK
Flybrid systems
Ford Motor Company
FreeScale Electronics
General Motors
GenRad
Google (Waymo)
haloIPT (Qualcomm)
Hella
HEVT
Honda

Hyundai
Institute of the Motor Industry (IMI)
Jaguar Cars
Kavlico
Ledder
Loctite
Lucas UK
LucasVarity
Mahle
Matlab/Simulink
Mazda
McLaren Electronic Systems
Mennekes
Mercedes
MIT
Mitsubishi
Most Corporation
NASA
NGK Plugs
Nissan
Nvidia
Oak Ridge National Labs
Peugeot
Philips
PicoTech/PicoScope
Pierburg
Pioneer Radio
Pixabay
Porsche

Renesas
Rolec
Rover Cars
Saab Media
SAE
Scandmec
Shutterstock
SMSC
Snap-on Tools
Society of Motor Manufacturers and Traders (SMMT)
Sofanou
Sun Electric
T&M Auto-Electrical
Tesla Motors
Texas Instruments
Thrust SSC Land Speed Team
Toyota
Tracker
Tula
Unipart Group
Valeo
Vauxhall
VDO Instruments
Volkswagen
Volvo Cars
Volvo Trucks
Wikimedia
ZF Servomatic

If I have used any information, or mentioned a company name that is not listed here, please accept my apologies and let me know so it can be rectified as soon as possible.

CHAPTER 1

Introduction

1.1 Why automated driving?

In this book I will cover all the key features of automated driving vehicles (ADVs) at a level that is ideal for technicians, but suitable for general consumption too. Many of the ideas picked up in this introduction chapter are examined in more detail later. The first important task is to understand why we need ADVs at all!

Definition

ADV: Automated driving vehicle

This industry commentator's view is similar to my own and many others:

> The economic and social benefits of autonomous driving [. . .] mean that it is inevitable. In fact, there may not be anything in history that provides benefits that [. . .] will dramatically reduce the cost of transportation for everyone. And it will enable children, [and] those with disabilities to have freedom to travel. It is difficult to find any change of technology that shows such vast benefits.

> *(McGrath 2018)*

Figure 1.1 Automated driving – on or off

It is believed, by most people in the industry, that connected and automated vehicles (CAVs) have the potential to:

▶ Reduce vehicle fatalities and injuries by as much as 90%
▶ Improve vehicle energy efficiency
▶ Reduce carbon emissions
▶ Reduce transportation costs
▶ Improve accessibility to transportation
▶ Cut the amount of land use for roads
▶ Reduce the need for parking

However, as Andreas Herrmann, Head of the Audi Lab for Market Research (amongst many other positions), states:

1 Introduction

"Despite all the euphoria about autonomous driving, the transition from driving oneself to being driven is still a considerable challenge both for customers and for car manufacturers. Can the pleasure of driving change into the pleasure of being driven?"

(Meyer-Hermann, Brenner, and Stadler 2018).

Also, research by the University of Michigan shows that despite significant investment in active safety systems, driver fatalities are increasing. Even if the increased miles that the vehicles travel is considered, driver fatalities are increasing, ending a long-term trend. Roadway departure is a key factor of this recent increase in fatalities. Figure 1.2 shows some of the trends in vehicle crashes.

In principle, an automated vehicle needs to have the same skills, or perhaps better skills, than a human driver:

▶ It has to be able to perceive and interpret its surroundings (Sense). For this, it uses the sensors just like we humans use our senses.
▶ The car needs to process information received and plan its driving strategy (Think). This task is undertaken by the vehicle computer using software and intelligent algorithms. This is done in a very short time and arguably faster in many cases than a human can process.
▶ It needs to use its powertrain, steering and braking system to move its wheels in such a way that the planned driving strategy is put into practice (Act).

Figure 1.3 is an infographic that explains this process pictorially.

Overall, the strongest argument in favour of automated driving is the reduction in accidents. However, there are many other issues to consider and we will examine these later in the book.

Key Fact

The main argument in favour of automated driving is the reduction in accidents.

For now, let's go back in time . . . I wrote the following article in 1993 as an introduction to my first textbook (*Automobile Electrical*

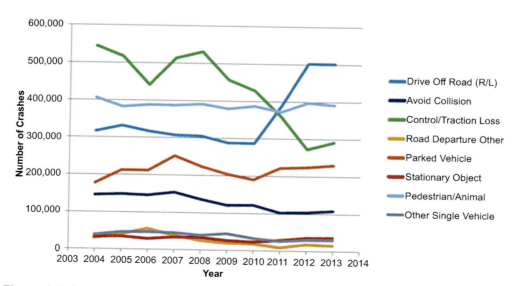

Figure 1.2 Single vehicle crashes. (Source: Crash Trends and Active Safety, University of Michigan, Carol Flannagan).

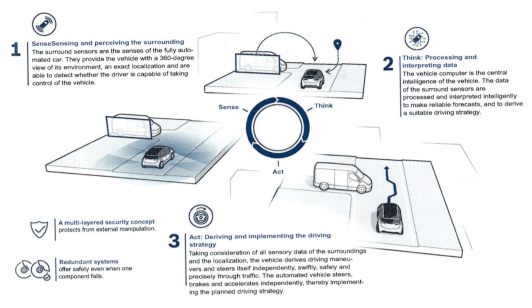

1 | **SenseSensing and perceiving the surrounding**
The surround sensors are the senses of the fully auto-mated car. They provide the vehicle with a 360-degree view of its environment, an exact localization and are able to detect whether the driver is capable of taking control of the vehicle.

2 | **Think: Processing and interpreting data**
The vehicle computer is the central intelligence of the vehicle. The data of the surround sensors are processed and interpreted intelligently to make reliable forecasts, and to derive a suitable driving strategy.

Sense   Think

Act

A multi-layered security concept protects from external manipulation.

Redundant systems offer safety even when one component fails.

3 | **Act: Deriving and implementing the driving strategy**
Taking consideration of all sensory data of the surroundings and the localization, the vehicle derives driving maneu-vers and steers itself independently, swiftly, safely and precisely through traffic. The automated vehicle steers, brakes and accelerates independently, thereby implement-ing the planned driving strategy.

Figure 1.3 Sense–Think–Act. (Source: Bosch Media).

and Electronic Systems). It seems I was quite good at predicting how automotive technology would develop:

Imagine what a vehicle will be like which is totally controlled by electronic systems. Imagine a vehicle with total on board diagnostic systems to pin-point any fault and the repairs required. Imagine a vehicle controlled by a 64 bit computer system with almost limitless memory. Imagine a vehicle with artificial intelligence to take all the operating decisions for you which also learns what you like and where you are likely to go. Finally imagine all of the above ideas combined with an automatic guidance system which works from cables laid under the road surface. Imagine what it would be like when it really went wrong!

However, picture this: Monday morning 15 January 2020, 08:00 hours. You are due at work by 09:00 which is just enough time to get there even though it is only fifteen miles away (the fourteen lane M25 soon filled to capacity), but

at least access to the wire guided lane helps.

A shiver of cold as you walk from the door of your house through the layer of snow makes you glad you paid the extra for the XYZ version of the 'car'. As you would expect the windows of the car are already defrosted and as you touch the thumb print recognition 'padd' and the door opens slowly a comforting waft of warm air hits you. It is still a little difficult to realise that the car anticipated that you would need it this morning and warmed the interior ready for your arrival.

Once the door is closed and the seat belts lift ready for you to snap into place a message appears on the windscreen. "Good morning Tom", you find that a little irritating as usual, "All systems are fully operational except the rear collision avoidance radar", (again), "I have taken the liberty of switching to a first line back-up system and have made a booking with the workshop computer via

3

the radio modem link". You can't help but feel some control has been lost, but still it's one less thing for you to worry about. "Shall we begin the journey, I have laid in a course for your work, is this correct?" Being able to speak to your car was odd at first but one soon gets used to these things. "Yes", you say, and the journey begins.

It is always comforting to know that the tyre pressures and treads automatically adjust to the road and weather conditions. Even the suspension and steering are fine tuned. The temperature as usual is now just right, without you even having to touch a control. This is because the temperature and climate control system soon learned that you prefer to feel very warm when you first sit in the car but like to cool the temperature down as the journey progresses. A small adjustment to the humidity would seem to be in order so you tell the car. "I will ensure I remember the change in future", appears on the screen.

Part way into the journey the car slows down and takes a turning not part of your usual route to work. The car decides to override the block you placed on audio communication, as it knows you will be wondering what happened. "Sorry about the change of route Tom but the road report transmission suggested this way would be quicker due to snow clearing." "We will still be at work on time".

The rest of the journey is uneventful and as usual you spend time working on some papers but can't resist seeing if you can hear when the diesel engine takes over from the electric. It's very difficult though because the active noise reduction is so good these days.

The car arrives at your place of work and parks in its usual place. For a change you remember to take the control unit with you, so the car doesn't have to remind you again. It's very good really though, as the car will not work without

it and you can use it to tell the car when you need it next and so on. The car can also contact you if for example unauthorised entry is attempted.

Finally, one touch on the outside control padd and the doors close and lock setting the alarm system at the same time.

While you are at work the car runs its fifth full diagnostic check of the day and finds no further faults. The sodium batteries need topping up, so the car sets a magnetic induction link with the underground transformer and the batteries are soon fully charged.

The car now drops into standby mode after having set the time to start preparing for your journey home which it has learnt has an 85% probability of being via the local pub . . . Prediction or science fiction?

(Denton 1995)

I got lots of things right but not everything!

Other commentators suggest that, "Some of the benefits of driverless cars will kick in only when most of the vehicular traffic on the road is fully autonomous. One such benefit would be the implementation of an automated traffic priority system that would ease commutes and help emergency vehicles more quickly ferry their ailing passengers to the hospital. Such a tiered traffic prioritisation system would require that all cars to be fully autonomous since human drivers don't always have the self-discipline or the big-picture view to follow route planning instructions appropriately."

(Lipson and Kurman 2018)

We will see what happens; however, in the meantime the section 1.2 will further define what is meant by autonomous or automated driving and the many other words and phrases that are in current use.

Figure 1.4 This Citroen DS19 from 1960 was used to experiment with a wire guided system in the UK

1.2 What is an automated driving vehicle (ADV)?

A fully autonomous car, also known as a driverless car, self-driving car and robotic car, is a vehicle capable of fulfilling the main transportation capabilities of a traditional car. It is capable of sensing its environment and navigating without human input. The vehicle does not have to be electrically powered, but almost all are (or will be). There are different levels of automation and these will be examined later.

Definition

Autonomous: having autonomy; not subject to control from outside; independent: a subsidiary that functions as an autonomous unit

Autonomous vehicle: navigated and manoeuvred by a computer without a need for human control or intervention under a range of driving situations and conditions (Source: www.dictionary.com)

For me, the word 'autonomous' implies that a vehicle has too much independence. For this reason, the word 'automated' works better in connection with most of these vehicles (level 5 is arguably autonomous, below that automated; see section 1.3). The technologies and associated systems are the same, but the consequences of how they are perceived could be an issue.

Thatcham Research[1] recently issued a warning about 'autonomous' vehicle marketing. This is because the word 'autonomous' in marketing materials produced by carmakers is lulling drivers into a false sense of security. Along with the Association of British Insurers (ABI), Thatcham Research has issued an urgent call to carmakers and legislators for greater clarity around the capabilities of vehicles sold with driver assistance technologies.

They suggest misleading names such as 'Autopilot' and 'ProPilot' are increasing the risks. These systems are designed to work, for example, on motorways, but can be used anywhere. Names like these do not help because they infer the car can do a lot more than it can.

James Dalton, Director of General Insurance Policy at the ABI, said:

> Given the part human error plays in the overwhelming majority of accidents, these technologies have the potential to dramatically improve road safety. However, we are a long way from fully autonomous cars which will be able to look after all parts of a journey.
>
> (Dalton 2019)

ADVs sense their surroundings with such techniques as radar, lidar, GPS and computer vision. Advanced control systems interpret sensory information to identify appropriate navigation paths, as well as obstacles and relevant signage. The vehicles are capable of updating their maps based on sensory input, allowing them to keep track of their position even when conditions change or when they enter unknown environments.

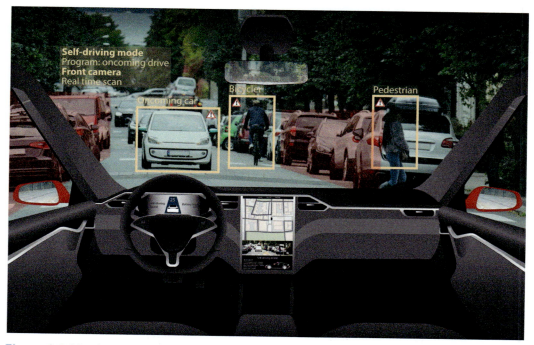

Figure 1.5 Live interpretation of a camera image

Figure 1.6 Audi piloted driving concept. (Source: Audi Media).

Key Fact

Lidar is a sensing technology that measures distance by illuminating a target with a laser and analysing the reflected light. The term was created as a portmanteau of 'light' and 'radar'.

The next section defines some useful terminology and also outlies the Society of Automotive Engineers (SAE) levels of automation. Vehicles at SAE levels 1 to 4

show increasing levels of automation and only at level 5 can they be described as autonomous. Others may think differently!

1.3 Definitions

1.3.1 Abbreviations

Learning about a new subject often involves new language and especially new abbreviations. Here are some that will be useful to refer back to from time to time:

▶ 5G – Fifth generation mobile wireless standard
▶ ADS – Automated driving system
▶ ADS-DV – Automated driving system – Dedicated vehicle
▶ ADV – Automated driving vehicle
▶ AEB – Automated Emergency Braking system – Detects vehicles and pedestrians in front and brings the vehicle to standstill during emergency

- Automated Valet Parking – Fully automated parking and summoning feature with no monitoring required. Can be executed with no passenger in the vehicle
- BSD – Blind Spot Detection – Warns the driver when the system detects other vehicles located to the driver's blind spot such side and rear
- Car sharing – The practice of sharing a car for regular travelling, especially for commuting
- CAV – Connected and autonomous vehicle
- Connected Car – A car which has technology enabling it to connect to devices within the car, as well as external networks such as the internet
- DDT – Dynamic driving task
- DDT fallback – Dynamic driving task fallback
- Emergency Driver Assistant – Automated assistance to steering and/or braking; acceleration functionality once the vehicle detects a delay in driver input for the driving situation
- Highway Assist – Steering and braking/ acceleration function automated during motorway; a road driving with driver required to monitor the operation
- Intersection Pilot – Automated lane merger and intersection driving during motorway; road driving with no monitoring from the driver required
- LDW – Lane Departure Warning – Warns the driver when the vehicle detects an unintentional drift from its travel lane
- LKA – Lane Keeping Assist System – Centres the vehicle to the middle of the lane when the vehicle detects an unintentional drift from its travel lane
- MRC – Minimum risk condition
- ODD – Operational design domain
- OEDR – Object and event detection and response
- OEM – Original Equipment Manufacturers of vehicles

- Parking Assist – Automated steering assistance to vehicle parking feature; driver needs to engage automated parking mode
- PDC – Park Distance Control – Ultrasonic sensor-based feature helping the driver in parking by providing audio warnings
- Ride hailing – Hailing a vehicle for immediate service; fees are usually time and distance-based
- Ride sharing – Passengers sharing vehicles where fees are often split between users
- Semi Assisted Valet Park – Steering and braking/acceleration feature automated in parking feature; driver needs to engage automated parking mode
- Traffic Jam Pilot – Automated low speed and stop and go driving with driver not needing to monitor the operation
- V2I (Vehicle-to-Infrastructure) – Technology which allows vehicle to communicate with infrastructures like parking garages, traffic signals etc.
- V2P (Vehicle-to-Pedestrian) – Technology which allows vehicle to communicate with pedestrians
- V2V (Vehicle-to-Vehicle) – Technology which allows vehicle to communicate with other vehicles

Figure 1.7 Driver monitoring camera. (Source: First Sensor AG).

- V2X (Vehicle-to-Everything) – Technology that allows vehicles to communicate with moving parts of the traffic system around them
- Valet Park Assist – Fully automated parking with driver monitoring the parking execution; driver may/may not need to engage automated parking mode

1.3.2 Descriptions

Here are some general descriptions relating to automated driving. They are used in particular in technical documents such as SAE papers.

- Operational design domain (ODD): The scenarios and circumstances under which the vehicle is designed to operate.
- Dynamic driving task (DDT): This includes the operational (steering, braking, accelerating, monitoring the vehicle and roadway) and tactical (responding to events, determining when to change lanes, turn, use signals, etc.) aspects of the driving task, but not the strategic (determining destinations and waypoints) aspect of the driving task.
- Driving mode: This is a type of driving scenario with characteristic dynamic driving task requirements (e.g., motorway merging, high speed cruising, low speed traffic jam, closed-campus operations, etc.).
- Request to intervene: This is notification by the automated driving system to a human driver that s/he should promptly begin or resume performance of the dynamic driving task.

1.3.3 SAE levels of driving automation

First issued in January 2014 and still being updated, SAE international's J3016 provides a common taxonomy and definitions for automated driving. This is to simplify communication and facilitate collaboration within technical and policy domains. It defines more than a dozen key terms and provides full descriptions and examples for each level (SAE 2018).

> **Definition**
> SAE: Society of Automotive Engineers www.sae.org

The SAE standard outlines six levels of driving automation that range from no automation to full automation. A key distinction is between level 2, where the human driver performs part of the dynamic driving task, and level 3, where the automated driving system performs the entire dynamic driving task.

> **Definition**
> Taxonomy: The practice and science of classification of things or concepts, including the principles that underlie such classification

SAE state that these levels are descriptive rather than normative, and technical rather than legal. They imply no particular order of market introduction. Elements indicate minimum rather than maximum system capabilities for each level. A particular vehicle may have multiple driving automation features such that it could operate at different levels depending upon the feature(s) that are engaged.

Figure 1.8 Self-driving racing car developed by Roborace

System refers to the driver assistance system, combination of driver assistance systems or automated driving system. Excluded are warning and momentary intervention systems, which do not automate any part of the dynamic driving task on a sustained basis and therefore do not change the human driver's role in performing the dynamic driving task (DDT).

Automation systems for on-road motor vehicles

The SAE has defined six levels of automation, ranging from no automation at level 0, to full autonomy at level 5.

Figure 1.9 outlines these levels in a less technical way:

Table 1.1 The levels of automation are defined from 0 to level 5

Level	Name	Definition	Dynamic driving task (DDT) Sustained lateral and longitudinal vehicle motion control	Dynamic driving task (DDT) Object and event detection and response (OEDR)	Dynamic driving task (DDT) fallback	Operational design domain (ODD)
0	No Driving Automation	The performance by the driver of the entire DDT, even when enhanced by active safety systems.	Driver	Driver	Driver	n/a
1	Driver Assistance	The sustained and ODD-specific execution by a driving automation system of either the lateral or the longitudinal vehicle motion control subtask of the DDT (but not both simultaneously) with the expectation that the driver performs the remainder of the DDT.	Driver and System	Driver	Driver	Limited
2	Partial Driving Automation	The sustained and ODD-specific execution by a driving automation system of both the lateral and longitudinal vehicle motion control subtasks of the DDT with the expectation that the driver completes the OEDR subtask and supervises the driving automation system.	**System**	Driver	Driver	Limited

(Continued)

Table 1.1 Continued

Level	Name	Definition	Dynamic driving task (DDT) Sustained lateral and longitudinal vehicle motion control	Dynamic driving task (DDT) Object and event detection and response (OEDR)	Dynamic driving task (DDT) fallback	Operational design domain (ODD)
3	**Conditional Driving Automation**	The sustained and ODD-specific performance by an ADS of the entire DDT with the expectation that the DDT fallback-ready user is receptive to ADS-issued requests to intervene, as well as to DDT performance-relevant system failures in other vehicle systems, and will respond appropriately.	System	**System**	Fallback-ready (the user becomes the driver during fallback)	Limited
4	**High Driving Automation**	The sustained and ODD-specific performance by an ADS of the entire DDT and DDT fallback without any expectation that a user will respond to a request to intervene.	System	System	**System**	Limited
5	**Full Driving Automation**	The sustained and unconditional (i.e., not ODD-specific) performance by an ADS of the entire DDT and DDT fallback without any expectation that a user will respond to a request to intervene.	System	System	System	**Unlimited**

Source: Copyright © 2018 SAE International. SAE International and J3016 are acknowledged as the source.

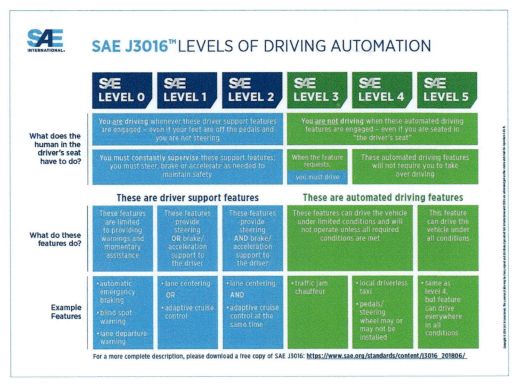

Figure 1.9 Simplified graphic showing levels of driving automation. (Source: SAE International J3016).

1.4 When will AVs become a common sight?

1.4.1 Introduction

In the UK, the Automated and Electric Vehicles Act 2018 became law on 19 July 2018:

The Automated and Electric Vehicles Act is intended to enable consumers in the United Kingdom to benefit from improvements in transport technology. The Act makes provision for (1) the creation of a new liability scheme for insurers in relation to automated vehicles, and (2) the creation of regulations relating to the installation and operation of charging points and hydrogen refuelling points for electric vehicles. The Act sets out the regulatory framework to enable new transport technology to be invented, designed, made and used in the United Kingdom.[2]

More details on this later.

A Department of Transport report in the UK called *The Pathway to Driverless Cars* determined that current UK legislation is not a barrier to their use and a Code of Practice was produced in 2015. The intention was that this would facilitate developments.[3]

In the USA, legislation was passed in several states (starting in 2012) allowing driverless cars. The number continues to increase.

In the Third Mobility Package, published on 17 May 2018, the European Commission included two significant measures: a proposal for a revision of the General Safety

11

Figure 1.10 Reading while the car drives

Regulation (GSR) and Pedestrian Safety Regulation (PSR) and a Communication (non-legislative) on Connected and Automated Mobility (CCAM). Since then, the European Commission has proposed a draft exemption to certification rules for automated vehicles and launched a roadmap for a future proposal on CCAM, while the European Parliament is negotiating a non-legislative response to the European Commission's proposals.

Some vehicle manufacturers have said that ADVs will start to become available in 2020, but the driving levels are not yet specified.

In 2016 those involved in driverless car research and development (R&D) in the UK suggested the cars would be on the roads in 2017. The Google car has done over 2 million miles, Tesla Autopilot more than 50 million miles (with one fatality as of 2018) and a company called Otto (part of Uber) shipped 5,000 bottles of beer in a self-driving lorry (so at least in the latter case the technology is proving useful). In Singapore, there are autonomous pods and driverless taxis in use. In the UK and many other countries, academia and industry are making great advances in the subject.

However, there are five major challenges that can hold back ADV developments; interestingly only one of these is wholly technology related:

▶ **Training**: Drivers are becoming used to some self-driving features, adaptive cruise control, lane guidance etc. They will need to further understand the limitations of the vehicles such that when the machine beeps, they need to take over.
▶ **Accidents**: In the event of a collision, it is not yet fully clear where the responsibility will lie. Decisions must be taken by the software, for example, to protect the car occupants first, or pedestrians first.
▶ **Stupidity**: Game theory researchers suggest that some people will learn to play chicken with the cars on crossings as they 'know' the car will have to stop. It may even be the case that more children play near roads because it is perceived to be a safer space. Animals are not stupid but when a bird or squirrel runs across the road – the car may decide to stop.
▶ **Interaction with people**: We have developed social norms to let people out

of junctions, or turn before us. Head nods, waves, light flashes etc. are used for this. The cars will 'talk' to each other but in a mix of autonomous and non-autonomous cars this is still to be solved.

▶ **Weather**: Cars cannot as yet manage a snowy road in the same way they can a nice bright sunny one. Road signs can become obscured, and slippery surfaces require a very different driving technique. Manufacturers are working on this, but there is still some way to go.

When will ADVs really start to become commonplace on our roads? My view is that while the level of automation on 'normal' vehicles will continue to increase, it will be something like 2030 before fully automated vehicles really hit the streets. There are still many problems to solve, not least of these the social ones – because they don't have a technical answer. All these issues will be looked at in more detail later.

Key Fact

It will be at least 2030 before fully automated vehicles start to be commonplace.

1.5 BMW view

Fully autonomous cars might never be allowed on many public roads, according to a top BMW executive. Speaking at the Society of Motor Manufacturers and Traders Summit, BMW's special representative to the UK, Ian Robertson, suggested that in the future cars will be highly intelligent but might never become truly self-driving in all conditions and on all roads.

BMW has a fleet of some 40 autonomous vehicles testing on public roads with each completing 1000 km journeys on a routine basis. However, Robertson has revealed that engineers still have to intervene on many trips.

He commented,

I believe that in the long term, the regulators will step in and set boundaries about how far we can go. It might be to allow it only on motorways, as they are the most controlled environments. Or perhaps they'd essentially 'rope off' parts of cities to allow autonomous cars into controlled areas, where the consequences for pedestrians are controlled.

Dr Robertson has also suggested that he could never envisage a scenario where BMW would make a car without a steering wheel, because he believes the driver would always want the option to be in control.

1.5.1 Tesla

Tesla was in the news again recently following founder Elon Musk's claim that its full self-driving feature would be complete by the end of this year, and would allow drivers to snooze in their seats by the end of 2020.

This ambition should be contrasted with the caution shown by most other car manufacturers, most of whom have been extending their timelines for the implementation of fully self-driving vehicles.

Can Tesla meet this latest self-driving promise? They often surprise us!

1.5.2 Volvo

Volvo has warned against premature self-driving roll-out. Hakan Samuelsson, the chief executive of Volvo Cars, warned that the premature launch of self-driving technology risks delaying "the best lifesaver in the history of the car."[4] According to the FT's Peter Campbell, the Volvo CEO said it was "irresponsible" to put autonomous vehicles on the road if they were not sufficiently safe, because doing so would erode trust among the public and regulators.

1.5.3 VW

In an interview with Reuters, the head of Volkswagen commercial vehicles warned that the cost and complexity of rolling out fully autonomous vehicle technology globally will serve to undermine the business case for doing so. Thomas Sedran told Reuters that he believes level 5 automation will never happen globally, given the need for the latest-generation mobile infrastructure, constantly updated high-definition digital maps and near perfect road markings. "The complexity of solving this problem is like a manned mission to Mars," he said.

Notes

1 The automotive research centre funded by the UK motor insurance industry
2 Source: https://services.parliament.uk/bills/2017-19/automatedandelectric vehicles.html
3 Source: www.gov.uk/government/publications/driverless-cars-in-the-uk-a-regulatory-review
4 Source: Financial Times (FT) 2019. https://www.ft.com/content/24d052b6-4cbb-11e9-8b7f-d49067e0f50d, accessed 24 March 19.

CHAPTER 2

Safety

2.1 Introduction

The basic operation, awareness and safe working practices relating to all vehicles is essential knowledge for anybody who works with them in any way. This can range from valeting, to sales and to complex diagnostics and repairs. All automated driving vehicles (ADVs) will also be electric vehicles (EVs), and there are serious safety risks if people are not trained and aware of the dangers in both these areas.

Safety issues associated with ADVs is a complex area and has to be considered by owners, repairers and the public at large. In addition, insurers and law enforcement will be affected by the changes that will come about as ADVs become more common.

Safety First

Do not operate or work on any vehicle that you do not understand or have not been trained on!

From the perspective of the vehicle, there are two areas that it must keep safe: itself and the occupants, and external people

Figure 2.1 ADVs will undergo extensive crash testing just like normal vehicles such as these

and property. Service and repair of these vehicles will be an area where there is also a significant risk – to the technician as well as other people if the technician gets something wrong.

2.2 Vehicle and its occupants

All the normal safety systems that we have come to expect in a modern vehicle will still be used in an ADV. These systems are usually described as being either active or passive. Active safety features engage to help prevent or reduce the severity of a crash. Passive safety features protect vehicle occupants during a crash.

Active safety systems only operate when needed. Forward collision warning systems and lane departure warning systems, for example, can activate a warning system when a dangerous situation is detected. Other systems like electronic stability control (ESC), anti-lock braking systems (ABS) and emergency brake assist (EBA) monitor the rotational speed of the vehicle's wheels, brake system operation and general vehicle stability, for any signs that control is being lost. When an issue is detected, these active safety features work autonomously to correct the situation.

Passive safety systems are those that help to protect vehicle occupants from injury during a crash. Their main function is to keep the vehicle occupants protected from the various crash forces. Passive safety features try to ensure that vehicle occupants remain in the inside space of the vehicle during the crash.

Crumple zones, for example, absorb some of the crash forces before they affect the occupants. Seatbelts, airbags and headrests help keep the people stationary within the vehicle. Passive safety features reduce the risk of serious injury. Arguably some passive systems are also active, but the key aspect is still that they don't do anything until a crash occurs!

Key Fact

Active safety systems only operate when needed.

Key Fact

Passive safety systems are those that help to protect vehicle occupants from injury during a crash.

Figure 2.2 Active safety components. (Source: Bosch Media).

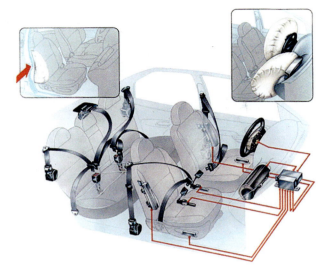

Figure 2.3 Passive safety components

2.3 External people and property

The safety of people and property (and animals) external to the vehicle is an area of great interest to designers of ADV systems. This is because the key difference from a human-driven vehicle is that now the machine will need to make decisions relating to the safety of external people and property, as well as the occupants and itself.

This brings up all the well-known scenarios about options such as choosing a head on crash rather than running into a group of children, or running over an animal to reduce the chance of a crash. Is the value of one life greater or lesser than another? More details about this difficult yet interesting area are presented later, together with how manufacturers are dealing with it.

2.4 Service and repair

This is an area where there is serious risk of injury to technicians. However, there is also a serious risk to car and occupants and external people if the work is not done

correctly – not doing a required software update, for example. All ADVs in the future will also be EVs, so there are the additional risks of high voltage and very strong magnets.

> **Safety First**
>
> Do not work on high voltage systems unless trained to the appropriate level

Safe working practices in relation to all automotive systems are essential, for your safety as well as that of others. When working with high voltage systems, it is even more important to be trained and qualified. There are some particular risks when working with electrical systems, but there are ways to reduce them. This is known as risk assessment.

> **Definition**
>
> Risk assessment: A systematic process of evaluating the potential risks that may be involved in an activity or undertaking

17

Figure 2.4 Personal protective equipment (PPE) in use on a high voltage system

Electric vehicles use high voltage batteries so that energy can be delivered to a drive motor or returned to a battery pack in a very short time. Voltages of 400V are now common and some are as much as 700V, so clearly, there are electrical safety issues when working with these vehicles. It is expected that the voltages will increase further in the near future on some vehicles.

Safety First

EV batteries and motors have high electrical and magnetic potential that can severely injure or kill if not handled correctly

EV batteries and motors have high electrical and magnetic potential that can severely injure or kill if not handled correctly. It is essential that you take note of all the warnings and recommended safety measures outlined by manufacturers. Any person with a heart pacemaker, for example, should not work on an EV motor since the magnetic effects could be dangerous. Other medical devices such as intravenous insulin injectors or meters can also be affected.

Other systems on ADVs are very complex and, if not serviced or repaired correctly, could put the vehicle in an unsafe condition. If a sensor or camera is not calibrated correctly, this could provide the vehicle's self-driving systems with incorrect data and result in unsafe operation. This is currently the case with existing advanced driver assistance systems but will become even more important as the levels of driving automation increase.

2.5 IMI TechSafe™

The Institute of the Motor Industry (IMI)[1] TechSafe is a campaign to direct employers to their health and safety responsibilities in relation to hybrid and electric, connected and automated driving vehicles. Employers have a legal duty of care under the Electricity at Work Regulations 1989 to ensure their employees are competent to work on the electrical systems in hybrid

Figure 2.5 IMI TechSafe logo

and electric vehicles. The EV Professional Standard, upon which the TechSafe campaign is based, aims to ensure that employers are working within the law, and that technicians remain skilled and competent over time. Technicians that meet the requirements of the scheme will be visible on a register of professionals, and able to use the TechSafe banner to show their mark of competence at different levels and in different subjects.

> **Definition**
>
> TechSafe: The only certain way to stay safe and prove your competence

The TechSafe campaign message is based on the EV Professional Standard: a model of training and registration of technicians based on EV qualifications, IMI Accreditation or accredited training, a code of professional behaviours, and a commitment to Continual Professional Development (CPD).

In the HSE's 'Electricity at Work: Safe Working Practices' (2013)[2] document, page 10 states:

> You must identify those people who are competent and have knowledge and experience of the electrical system to be worked on. Anyone who does not have this will need a greater level of supervision, or will need to be given adequate training to make sure that they have the correct skills, knowledge and risk awareness for the task. Do not let unauthorised, unqualified or untrained people work on electrical systems.

IMI TechSafe™ registration is the ideal way for an employer, firstly, to prevent danger or injury, and secondly, to be the defence in case of criminal proceedings.

Notes

1 The Institute of the Motor Industry: www.theimi.org.uk
2 Source: www.hse.gov.uk/pubns/books/hsg85.htm

CHAPTER 3

Advanced driver assistance systems (ADAS)

3.1 Introduction

Advanced driver assistance systems (ADAS) are, as the name suggests, designed to help the driver. This improves safety because most road accidents occur due to human error. Automated systems help to minimise human error, which has been proven to reduce road fatalities. These are also the enabling technologies for full automated driving.

> **Key Fact**
>
> Automated systems help to minimise human error.

As an interesting example to show the benefits of ADAS:

There was an incident December 2016 when a Tesla saved a driver by suddenly activating the collision warning, and the autopilot turned on the emergency brake. It was not until after this manoeuvre that a car in front of the driver flipped and landed in the path of the oncoming Tesla. Apparently, the sensors could tell that a threat had developed two cars ahead of the Tesla, which is why the driver did not see it. In other words, an accident was prevented because the radar was able to detect a situation beyond the driver's field of vision.

(Meyer-Hermann, Brenner, and Stadler 2018)

ADAS relies on inputs from several sources such as lidar, radar, cameras and vehicle CAN data. Safety features are designed to alert the driver to potential problems, or to avoid collisions by implementing safeguards. In some cases, this means taking control of the vehicle.

Figure 3.1 Sensing activity

Figure 3.2 Automatic parking. (Source: Park4U).

Key Fact

ADAS relies on inputs from several sources.

ADAS features can, for example, switch on lights, provide adaptive cruise control and collision avoidance, incorporate traffic warnings, alert the driver to other vehicles and dangers, warn if lane departure is detected or even initiate automated lane guidance. Cameras are used to see in blind spots.

Many modern vehicles now have systems such as electronic stability control, anti-lock brakes, lane departure warning, adaptive cruise control and traction control. All these systems can be affected by mechanical alignment. For this reason, correct repairs, adjustments and servicing are essential.

3.2 Example systems

Some example advanced driver assistance systems are listed here:

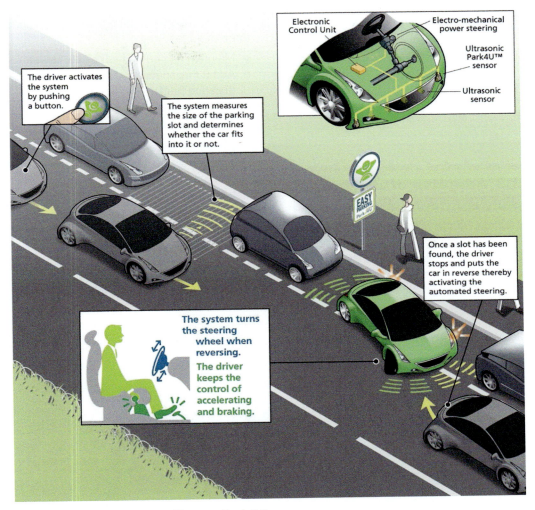

Electronic Control Unit

Electro-mechanical power steering

Ultrasonic Park4U™ sensor

Ultrasonic sensor

The driver activates the system by pushing a button.

The system measures the size of the parking slot and determines whether the car fits into it or not.

Once a slot has been found, the driver stops and puts the car in reverse thereby activating the automated steering.

The system turns the steering wheel when reversing.

The driver keeps the control of accelerating and braking.

Figure 3.3 Parallel parking. (Source: Park4U).

▶ Adaptive cruise control
▶ Adaptive light control
▶ Anti-lock braking system
▶ Automatic parking
▶ Blind spot monitor
▶ Collision avoidance
▶ Collision warning
▶ Driver drowsiness detection
▶ Electric vehicle warning sounds
▶ Emergency driver assistant
▶ Glare-free high beam and pixel light
▶ Lane change assistance
▶ Lane departure warning

Figure 3.4 Camera used for night vision. (Source: First Sensor AG).

23

- ▶ Navigation system with traffic information
- ▶ Night vision
- ▶ Parking sensor
- ▶ Pedestrian protection
- ▶ Rain sensor
- ▶ Surround view
- ▶ Traffic sign recognition
- ▶ Tyre pressure monitoring

3.3 Adaptive cruise control

Conventional cruise control is not always practical on many European roads. This is because the speed of the general traffic varies constantly, and traffic is often very heavy. The driver must take over from a standard cruise control system on many occasions to speed up or slow down. Adaptive cruise control (ACC) can automatically adjust the vehicle speed to the current traffic situation. The system has three main features:

- ▶ Maintain a speed as set by the driver
- ▶ Adapt this speed and maintain a safe distance from the vehicles in front
- ▶ Provide a warning if there is a risk of collision.

The main extra components, compared to standard cruise control, are the headway sensor and the steering angle sensor; the first of these is clearly the most important. Information on steering angle is used to further enhance the data from the headway sensor by allowing greater discrimination between hazards and spurious signals. Two types of headway sensor are in use: radar and lidar. Both contain transmitter and receiver units. The radar system uses microwave signals of up to 80 GHz, and the reflection time of these gives the distance

Figure 3.5 Adaptive cruise control

to the object in front. Lidar uses a laser diode to produce infrared light signals, the reflections of which are detected by a photodiode.

Key Fact

Radar system uses microwave signals of up to 80 GHz, and the reflection time of these gives the distance to the object.

These two types of sensors have advantages and disadvantages. The radar system is not affected by rain and fog, but the lidar can be more selective by recognising the standard reflectors on the rear of the vehicle in front. Radar can produce strong reflections from bridges, trees, posts and other normal roadside items. It can also suffer loss of signal return due to multipath reflections. Under

Figure 3.6 Lidar sensor

Figure 3.7 Headway sensor is fitted at the front of a vehicle. (Source: Bosch Media).

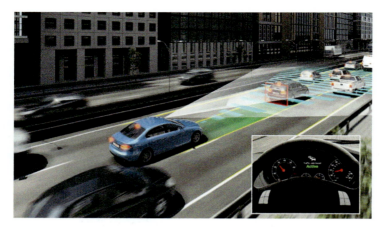

Figure 3.8 Forward sensing

Figure 3.9 Reversing aid as part of a control system. (Source: Ford).

ideal weather conditions, the lidar system appears to be the best, but it becomes very unreliable when the weather changes. A beam divergence of about 2.5° vertically and horizontally has been found to be the most suitable whatever headway sensor is used. An important consideration is that signals from other vehicles fitted with this system must not produce erroneous results. Figure 3.7 shows a typical headway sensor and control electronics.

Fundamentally, the operation of an adaptive cruise system is the same as a conventional system, except when a signal from the headway sensor detects an obstruction. In this case, the vehicle speed is decreased. If the optimum stopping distance cannot be achieved by just backing off the throttle, a warning is supplied to the driver. Later systems also take control of the vehicle transmission and brakes.

3.4 Obstacle avoidance radar

This system, sometimes called collision avoidance radar, can be looked at in two ways. First, as an aid to reversing, which gives the driver some indication as to how

Figure 3.10 Rear sensors

Figure 3.11 Audi ultrasonic rear sensor

3.5 Basic reversing aid

A reverse sensing system is a reverse only parking aid system that uses sensors mounted in the rear bumper. Parking aid systems feature both front and rear sensors. Low-cost, high-performance ultrasonic range sensors are fitted to the vehicle. Generally, four sensors are used to form a detection zone as wide as the vehicle. A microprocessor monitors the sensors and emits audible beeps during slow reverse parking to help the driver reverse or park the vehicle. The technique is relatively simple as the level of discrimination required is low and the system only has to operate over short distances.

much space is behind the car. Second, it can be used as a vision enhancement system.

The principle of radar as a reversing aid is illustrated in Figure 3.9. This technique is, in effect, a range-finding system. The output can be audio or visual, the latter being perhaps most appropriate, as the driver is likely to be looking backwards. The audible signal is a 'pip-pip-pip' type sound, the repetition frequency of which increases as the car comes nearer to the obstruction and becomes almost continuous as impact is imminent. Many systems now also make the noise come from the appropriate speaker(s) to indicate direction.

3.6 Radar

Figure 3.12 is a block diagram to demonstrate the principle of a radar system. A frequency of 79 GHz with a bandwidth of 4 GHz is generally used and this has a resolution of 0.1 m. 77 GHz systems are still also in common use.

27

3 Advanced driver assistance systems (ADAS)

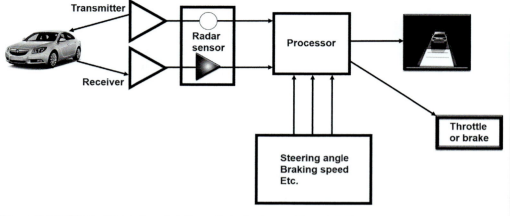

Figure 3.12 Block diagram animation of obstacle avoidance radar

The operation of a basic radar system is as follows: a radio transmitter generates radio wave pulses, which are then radiated from an antenna. A target, such as another vehicle, scatters a small portion of the radio energy back to a receiving antenna. This weak signal is amplified and displayed on a screen. To determine its position, the distance (range) and bearing must be measured. Because radio waves travel at the speed of light $(3 \times 10^8$ m/s), the range may be found by measuring the time taken for a radio wave to travel from transmitter to obstacle, and back to the receiver. For example, if the range were 150 m, the time for the round trip would be:

$$t = \frac{2d}{C}$$

Where: t = time, d = distance to object, and C = speed of light.

In this example it is 1 microsecond.

$$t = \frac{2 \times 150}{3 \times 10^8} = 1 \mu S$$

In other words, if the measured time for the round trip is 1 microsecond then the distance must be 150 m. Relative closing speed can be calculated from the current vehicle speed. The type of display or output that may be used on a motor vehicle will vary from an audible warning to a warning light or series of lights, or more likely a display screen.

3.7 Stereo video camera

Emergency braking systems are among the most effective assistance systems in the car. It is estimated that something like 70% of all rear-end collisions, resulting in personal injury, could be avoided if all vehicles were equipped with them. Bosch has developed a stereo video camera such that an emergency braking system can function based solely on camera data. Normally, this would require a radar

Figure 3.13 Land Rover using the camera system. (Source: Bosch Media).

Figure 3.14 Stereo video camera for ADAS. (Source: Bosch Media).

sensor or a combination of radar and video sensors.

Land Rover, for example, uses a stereo video camera together with the Bosch emergency braking system as standard in its Discovery Sport. When the camera recognises another vehicle ahead in the lane as an obstruction, the emergency braking system prepares for action. If the driver does not react, then the system initiates maximum braking.

Other driver assistance functions can also be based on the stereo video camera. One such function is road-sign recognition, which keeps the driver informed about the current speed limit. Another is a lane-departure warning. This vibrates the steering wheel to warn drivers before they unintentionally drift out of lane.

With its light-sensitive lenses and video sensors, the camera covers a 50° horizontal field of vision and can take measurements in 3D at over 50 m. Thanks to these spatial measurements, the video signal alone provides enough data to calculate, for example, the distance to vehicles ahead. Its pair of highly sensitive video sensors are equipped with colour recognition and CMOS (complementary metal oxide semiconductor) technology. They have a resolution of 1,280

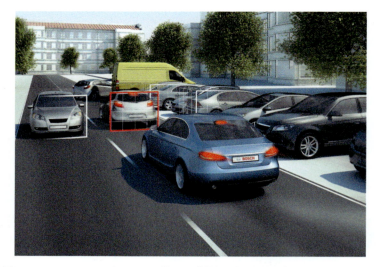

Figure 3.15 The camera recognises another vehicle ahead in the same lane. (Source: Bosch Media).

by 960 pixels and can also process high-contrast images.

3.8 Rear radar

Drivers are taught to assess surrounding traffic, before changing lanes, by checking their rear-view and side mirrors and looking over each shoulder. However, the area alongside and just behind the vehicle is a constant source of danger and often the cause of serious accidents. Drivers are not able to see into this area using either the rear view or side mirrors, but it is big enough for a vehicle to be missed by a cursory glance.

To help minimise this risk, a lane-changing assistant receives the information it needs from a mid-range radar sensor for rear-end applications. This means drivers are effectively looking over their shoulders all the time, because it reliably and accurately recognises other road users in their vehicle's blind spot.

A typical installation is to have two sensors in the rear bumper, one on the left, one on the right. These two rear sensors monitor the area alongside and behind the car. Powerful control software collates the sensor information to produce a complete picture of all traffic in the area behind the vehicle. Whenever another vehicle approaches at speed from behind or is already present in the blind spot, a signal such as a warning light in the side mirror alerts the driver to the hazard. Should the driver still activate the turn signal with the intention of changing lanes, the lane-changing assistant issues an additional acoustic and/or haptic warning.

The rear radar system can do more than just assist with lane-changing. These sensors also form part of a cross-traffic alert system, which supports drivers reversing out of perpendicular parking spaces when their rear view is obstructed. Able to recognise cars, cyclists and pedestrians crossing behind the reversing vehicle at up to 50 m, the system alerts the driver to the imminent danger of collision by issuing an audible or visible signal.

Figure 3.16 Sensors monitor all traffic in the area behind the vehicle

Figure 3.17 Mid-range radar (MRR) sensor. (Source: Bosch Media).

Key Fact

Advanced radar systems can recognise cars, cyclists and pedestrians behind the vehicle at up to 50 m.

The Bosch mid-range radar sensor (MRR) is a bistatic multi-mode radar with four independent receiver channels and digital beam forming (DBF). It operates in the 76–77 GHz band that is standard for automotive radar applications in almost all countries worldwide.

A newer development known as an MMR sensor has an aperture angle of up to 150° and a range of up to 90 m. The forward-facing version looks significantly further; with an aperture angle of up to plus/minus 45°, it can detect objects up to 160 m.

There is also a long range radar (LRR) sensor which is a monostatic multimodal radar that has six fixed radar antennae. The central four antennae feature optimum properties for recording the vehicle's surroundings at higher speeds. They create a focused beam pattern with an opening angle of ±6 degrees, providing excellent long range detection with minimal interference from traffic in adjacent lanes.

Short range radar (SRR) works up to 30 m, medium range radar (MRR) up to about 100 m, and long range radar (LRR) up to around 200 m. The beam angle is narrower at longer ranges.

Definition

SRR: Short range radar
MRR: Medium range radar
LRR: Long range radar

3.9 Functional safety and risk

Functional safety depends on the system or equipment operating correctly in response to a range of inputs. This includes management of likely operator errors, hardware failures and changing environments. It describes the overall requirement for fault tolerance in a system or piece of equipment. The aim is no unacceptable risks of physical injury, and no damage to the health of people. Figure 3.18 illustrates the concept as a diagram.

Functional safety is a specific aspect of safety management, which relates to a particular system or piece of equipment operating correctly. It looks at the potential dangers arising from system or piece of equipment malfunctioning.

Definition

Functional safety is a specific aspect of safety management, which relates to a particular system or piece of equipment operating correctly

There are two main considerations relating to safety in cars:

▶ reducing dependency on driver skill and awareness
▶ functional safety.

Driver support systems such as automatic emergency braking, lane-departure warning and pedestrian detection are helping to reduce the number of accidents that have a human causal element. According to the National Highway Traffic Safety Administration (NHTSA) in the USA, the figure is currently around 94%.

Definition

NHTSA: The National Highway Traffic Safety Administration is an agency of the Executive Branch of the U.S. government, part of the Department of Transportation.

Because of the move towards fully autonomous driving as part of a fully-integrated transport system, the need for 'at the edge' processing will increase by orders

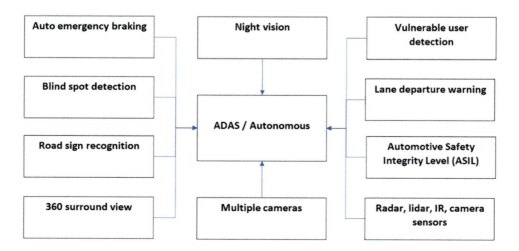

Figure 3.18 ADAS functional safety diagram

of magnitude. It is suggested by some industry insiders that the extra processing power needed for this can be done in the cloud. There are two issues with this; first, an always-on connection would be required, and even 5G will not deliver this at least in the shorter term. Secondly, a car must be able to operate completely offline.

In fully-autonomous cars, there will be a massive need for AI to make sense of what the cameras and other sensors 'see'. It is estimated that up to 20 cameras could be used.

All this additional electronic functionality, which is primarily to reduce driver error, means there is a need to have a functional-safety approach across the design process of the entire system. This is not a test of the actual function of a device or system, but whether it has integrity.

The industry is undergoing a fundamental change in car-electronics architecture. The car is actually becoming more like a software platform. Centralised computing functions connected by gigabit Ethernet will allow a whole new set of Application Programming Interfaces (APIs). These in turn will support the delivery of new features and services.

If we don't get the safety aspects of these new systems right, there is a danger that instead of reducing the human errors that cause accidents, the new electronic systems will simply cause them instead!

Safety First

Safety aspects of ADV systems must be right or they could cause more accidents not less.

Key Fact

A functional-safety approach across the design process is not a test of the actual function of a device or system, but whether it has integrity.

CHAPTER 4

Automated driving technologies

4.1 How does a human drive?

4.1.1 Introduction

There are arguably an infinite number of situations that a human driver has to deal with when operating a vehicle. All of them involve the process of sensing, understanding and then acting. See, think, act is another popular version of the process, but they all mean the same thing.

Turns across traffic account for about 20% of accidents, turns the other way only for about 1% (McGrath 2018), so we will look at this manoeuvre or action in more detail

to illustrate the challenge faced by humans and machines.

Key Fact

Turns across traffic account for about 20% of accidents.

Example situation: driving along the road at the normal speed and then turning right (UK, Japan, Australia etc.) or turning left (Europe, North and South America etc.) from a normal two-lane, two-way road into another two-lane, two-way road at right angles. Figure 4.2 shows the steps a human will take:

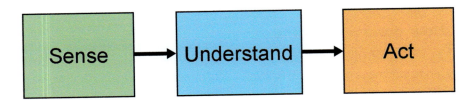

Figure 4.1 Sense, understand, act

4 Automated driving technologies

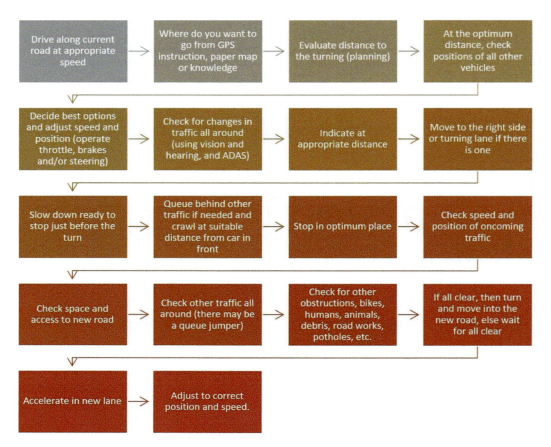

Figure 4.2 Sequence of events when turning across oncoming traffic

I am sure I have missed something in the above sequence (answers on a postcard please) but this illustrates the complexity and perhaps why learning to drive is difficult. Once you are a practiced driver many of the points on the list become almost instinctual. Not for a robot, though; it has to calculate each step and more, even if it can learn new methods over time.

4.1.2 How does a machine drive?

Arguably, an autonomous car (level 5) follows the same process as the human and can even get itself out of a problem situation if necessary. Automated driving vehicles (ADVs;

levels 1 to 4) do similar things but at lower levels, and they require human intervention for problems they cannot understand.

> **Definition**
> BASIC: Stands for Beginners All-purpose Symbolic Instruction Code, but is now a quite advanced computer programming language.

Arguably, every single action that an ADV takes will follow some simple (BASIC) computer code, which could be something like this (*very* much simplified!) example. In this case it is working to keep a constant speed at up to a set value:

```
If road_ahead = clear And speed < current_set_value Then
accelerate = True
brake = False
driver_message = "Ok"
ElseIf road_ahead <> clear Or speed > current_set_value Then '<> means doesn't equal
accelerate = False
brake = True
driver_message = "Caution"
End If
```

This action would be repeated over and over at very high speeds. In reality the process is far more complex, but the principle is the same.

Interpreting sensor input data is the most processor hungry operation. To set a context for what processing speed is needed, consider this: a vehicle travelling at 100 km/h (about 62 mph) is covering about 28 m/s. This means that in just 10 ms (milliseconds) the car will still move 28 cm. Additional details of the calculation process and complexity are represented by Figure 4.4. The need for very fast processors is clear when we examine path planning in the next section.

Key Fact

A car moving at 100 km/h is about 28 m/s and in 10 ms will move 28 cm.

Because of the complexity of automated driving a huge quantity of data is generated. When an A380 Airbus flies from London to New York in autopilot mode it requires 2.5

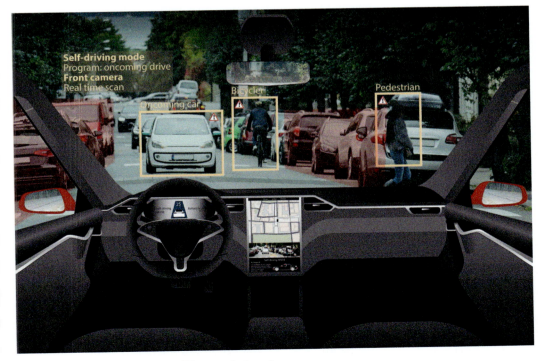

Figure 4.3 Processed view of the road ahead

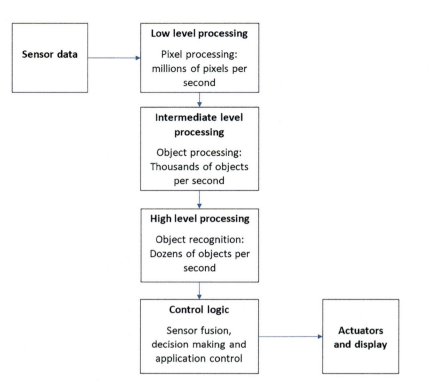

Figure 4.4 Handling increased system complexity in ADAS and ADV applications

MB of data. A level 4 vehicle can require 45 TB of data just to handle routine tasks.

4.2 The road to autonomy

4.2.1 Introduction

To rise through the levels from driver assistance, through automation and finally autonomy, is a difficult road for a car to navigate. In fact, roads are difficult for cars to navigate. Whilst far from exhaustive, the following short sections outline the basic things that an autonomous vehicle must be able to do. As experts in this area note, this is not easy:

When compared to air traffic, road traffic is a more chaotic system, albeit one that constantly reorganises itself. There are fundamental rules of the road, but the wide diversity of traffic situations cannot be regulated in all of their many complex details.

(Meyer-Hermann, Brenner, and Stadler 2018)

Until we reach level 5 autonomy the driver needs to be in a position to take control, so for example, must not be asleep! Let's now examine what is necessary for an ADV to actually control a vehicle on the roads.

> **Key Fact**
>
> Until level 5 autonomy is achieved, the driver needs to be in a position to take control.

Figure 4.5 Driving scenarios. (Source: Toyota).

4.2.2 Path planning

The simple aim of path planning is getting the vehicle to its destination. The actual sensors and methods used by different manufacturers will vary a little, but the algorithm used for path planning will be something like the following description.

The vehicle needs to determine an approximate long range path, which is made up of constantly changing short range paths. These short range paths (turn left, speed up, change lanes etc.) are what the vehicle is capable of achieving under the current operating conditions. Any short range paths that involve coming too close to an obstacle, entering the path of an oncoming vehicle without enough time to complete the manoeuvre and similar paths are rejected as options. Another example is a vehicle travelling at 90 kph (25 m/s), which would not be able to turn sharp left in 25 m, but it would be able to take a shallow curve in the road, without changing speed, if it was clear for a suitable distance ahead.

> **Key Fact**
> Any short range paths that are dangerous are rejected as options.

The remaining feasible paths are evaluated, and once the best path is determined, a set of throttle, brake and steering commands are sent to the appropriate system controllers and actuators.

> **Key Fact**
> A long range path is made up of constantly changing short range paths.

The time required for this 'decision' varies depending on the system's architecture and processor speed but is in the order of 50 milliseconds (a twentieth of a second). This whole process is repeated many times until the car reaches its destination. To add a context here, at 90 kph (25 m/s) a car would travel about 1.25 m, and the short

39

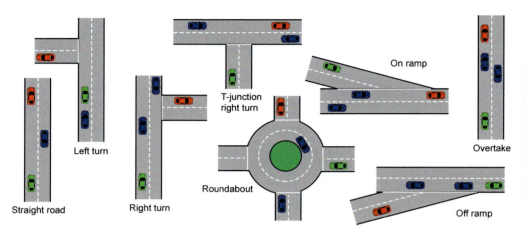

Figure 4.6 Path planning strategy and evolution

range path planning process would repeat – arguably much faster than a human could react.

There is a range of actions that a vehicle may need to carry out. Some of these are represented (driving on the left) in Figure 4.6. In each case the green car is the current position and the red car represents the position after the chosen action has completed. The blue car, or cars, represent typical traffic hazards. In reality, there will be even more complexity, for example pedestrians, parked cars, road works and much more.

4.2.3 Turning a corner

Turning a corner is clearly an essential function for an autonomous vehicle. In an ideal world, this could all be planned by using GPS data, but currently this is not accurate enough. The data also needs to be fully up to date. Having just completed a journey where my satellite navigation (GPS) informed me I was in a field while I was actually doing 70 mph on a motorway, I can confirm that this can be an issue!

For precise positioning high definition (HD) mapping is necessary. With this system the vehicle would know where it was within

about 15 cm. Turning a corner, therefore, is well within the vehicle's ability, assuming it is always looking for feasible paths as previously mentioned. Vehicles that use lidar towers are constantly updating their own 3D HD map.

Figure 4.7 GPS map. (Source: Google Maps).

position of all obstacles within a certain range. This is a continuous process as part of the short range path planning procedure.

The most difficult corner to turn is one that cuts across oncoming traffic as was discussed previously for a human driver.

4.2.4 Turning across oncoming traffic

Turning across oncoming traffic is a complex process. Whether it is a left or right turn of course changes depending on what side of the road is used. For this example, we will assume traffic lights are not used.

Estimating the speed of oncoming traffic is essential for this manoeuvre. However, on a very busy road it is possible that the car will never be able to complete the turn if the speed and distance of the oncoming vehicle does not allow a short-term path to put on the feasible list – or at best it may wait 15 minutes before suitable conditions occur.

A human driver would probably decide after a while that they will make the turn in a smaller gap than they would normally, or maybe if an oncoming driver flashed their lights.[1] They could even assume that the oncoming driver would accept or understand the situation and slow down a little when they saw your car moving across.

If traffic lights are used, then the oncoming traffic is not as much of an issue, but the car will need to determine if it can complete the turn before the lights change. This is a situation where a connected car would ideally be able to access traffic light timings.

4.2.5 Obstacle avoidance

An automated or autonomous vehicle (AV) needs to identify the current and predicted

Obstacles are categorised using a library of pre-determined shapes and motion descriptions. This library can be updated as the vehicle learns. The previous, current and predicted positions of moving objects are stored in the internal map. The shape, as well as the speed of an object, helps the vehicle choose the correct category. A two-wheeled vehicle moving at 50 mph is more likely to be a motorcycle than a bicycle for example. The vehicle can now plan a path that avoids obstacles.

4.2.6 Collision avoidance and evasion

Collision avoidance can also become collision mitigation, in other words the *least-worst* crash scenario is selected! Many ADAS vehicles already have collision avoidance but they only warn the driver and pre-charge or possibly apply the brakes, they do not mitigate.

Front-facing sensors such as cameras, radar or lidar are used and often form part of an active cruise control (ACC) system. This is SAE automated driving level 1.

Autonomous systems need to decide if collision is going to occur, and if corrective or evasive action is not possible, the type of crash must be chosen. This is arguably one of the most contentious areas of autonomous technology: what does a car decide if the options are head on high speed crash that will kill its occupants, run over a pedestrian or crash into a static object. Statements from some manufacturers have suggested that the car will always look after its occupants. More on this in Chapter 5.

Figures 4.8, 4.9 and 4.10 illustrate the Sense–Think–Act process for avoiding the car in front by overtaking it (this example situation is in a right hand lane driving region).

Key Fact

Statements from some manufacturers have suggested that the car will always favour its occupants in an accident.

Definition

Sense–Think–Act: This is how a human reacts to a stimulus; it is the same with an ADV.

Figure 4.8 Sense

Figure 4.9 Think (understand)

Figure 4.10 Act

Figure 4.11 The complexity of even a relatively quiet road

4.2.7 Traffic sign recognition

Traffic sign recognition (TSR) was first used in 2008/9 on a BMW; it recognised speed limit signs. Many systems now in use recognise 'speed limits' but also 'children', 'turn ahead' and similar signs. Fortunately, the Vienna Convention on Road Signs and Signals (1968) standardised signs. This means that using a camera and a suitable processing system, it is possible to recognise them across many different countries.

Definition

TSR: Traffic sign recognition

Signs that have been damaged due to a road accident or inclement weather could however be a problem.

4.2.8 Traffic light detection

Cameras are necessary for traffic light detection (TLD). They need to recognise

43

the units and also the colours (red, amber, green). Various outdoor conditions, obstructions and damage can make recognising lights difficult. However, solutions that are using a map the car will 'know' when to expect a traffic light and in the future, it is likely that the lights will communicate with the cars by 5G or similar. Stopping on red and going on green must be 100% accurate.

> **Definition**
>
> TLD: Traffic light detection

4.2.9 Emergency vehicle detection

An emergency vehicle can be detected first by sound and then by cameras so that the approaching path can be determined. Actions such as pulling off the side of the road are a challenge as the condition of the surface could be an issue.

Ultimately, it is expected that all emergency vehicles will communicate with the 'network' so that connected cars will know when they are approaching and can take early evasive action.

4.2.10 Other situations

No matter how much planning and preparation goes into a fully autonomous vehicle (level 5), there will always be situations that are unique. The vehicle will need to be able to deal with them or re-route around them.

> **Key Fact**
>
> There will always be situations that are unique; a fully autonomous vehicle (level 5) vehicle will need to be able to deal with them or re-route around them.

Pedestrians are one of the most unpredictable things we meet when driving! Even a simple situation like a pedestrian waiting to cross the road when not at a designated crossing place can be a challenge. For example, a driver may wave to the pedestrian or flash their lights to say they are going to stop or slow down. A pedestrian may stumble by the side for the road and could look like they are about to cross. They could even decide to play chicken with an autonomous car to make it stop!

Roundabouts can be a particular challenge. Some now consist of smaller roundabouts around a larger one. Once local drivers understand these it is possible to choose the best route that can appear to be going the wrong way around the big central circle but is actually following roundabout-rules for all the smaller ones. Some roundabouts are also so busy that at first view it is impossible to get on them, and then very difficult to get off.

Road surfaces such as cobblestones or blocks may be a particular issue for autonomous cars. This is partly because the surface is unpredictable, but also because in many places there are no white lines on them to preserve their historical value.

I live in the countryside in the UK, and there are lots of narrow roads (like Figure 4.12) with passing places, or in many cases both cars have to pass slowly with the nearside wheels off the road. Several years ago some friends from a large USA city visited, and as I was driving with them as passengers on these roads they asked, "What happens if a car comes the other way?" I had not given the problem much thought, just assuming that I or the oncoming car would move over, or in some extreme cases, one of us would have to reverse to a passing place. Their comments, however, illustrate the challenge for a fully autonomous car – particularly if two of them

Figure 4.12 Narrow roads. (Source: Stephen McKay, CC).

meet on this type of narrow road. Who has to reverse? Will there be a hierarchy where the newer car gets priority?

4.3 Perception

4.3.1 Introduction

In order to deal with all the situations outlined in the previous sections, ADVs need to sense or perceive their surroundings. Perception is a word that has a number of meanings, for example the:

- ability to see, hear or become aware of something;
- awareness of something through the senses; or
- way in which something is understood, regarded or interpreted.

These definitions generally relate to human perception, but a machine can perceive its surroundings in similar ways. The machine in this case is an autonomous (automated, robotic) vehicle (car). The car perceives its surroundings through a number of sensors rather than senses; the information from these sensors is processed, after which we can say that it has perception. Figure 4.13 shows a vehicle with a typical range of sensors.

The main sensors used by ADVs are:

- Cameras
- Radar
- Lidar
- Ultrasonic
- GPS
- Microphones (sound)

These are similar to the ADAS sensors outlined earlier. In addition to these sensors, V2X and cloud information can be used, for example, for communication about a traffic jam ahead or the vehicle in front providing information on the gap from it to the car in front. More about this later. Some designs use all of these sensors, some use a different combination.

Definition

V2X: Vehicle (connected) to everything

45

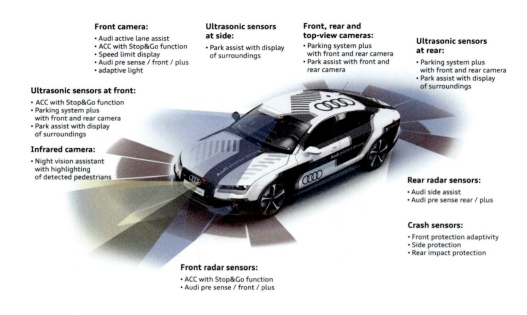

Front camera:
- Audi active lane assist
- ACC with Stop&Go function
- Speed limit display
- Audi pre sense / front / plus
- adaptive light

Ultrasonic sensors at side:
- Park assist with display of surroundings

Front, rear and top-view cameras:
- Parking system plus with front and rear camera
- Park assist with front and rear camera

Ultrasonic sensors at rear:
- Parking system plus with front and rear camera
- Park assist with display of surroundings

Ultrasonic sensors at front:
- ACC with Stop&Go function
- Parking system plus with front and rear camera
- Park assist with display of surroundings

Infrared camera:
- Night vision assistant with highlighting of detected pedestrians

Rear radar sensors:
- Audi side assist
- Audi pre sense rear / plus

Crash sensors:
- Front protection adaptivity
- Side protection
- Rear impact protection

Front radar sensors:
- ACC with Stop&Go function
- Audi pre sense / front / plus

Figure 4.13 Vehicle components and systems on an Audi RS 7 showing the piloted driving concept. (Source: Audi Media).

4.3.2 Cameras

Cameras are essential for object detection. They supply the vehicle with the data so that it can use AI to detect objects, such as other vehicles, pedestrians and road signs. Cameras can measure angles precisely so that, for example, the vehicle can determine if an approaching vehicle is likely to turn.

A wide field of view is for smaller roads in towns and up to 300 meters with a narrow angle of vision are ideal on major roads. Cameras will also 'see' the lane markings so a lane keeping assist function is possible.

Tesla insist they can do full autonomy with only cameras but usually they are combined with other sensors such as radar and lidar.

4.3.3 Radar

Radar systems are an active technology. They emit electromagnetic waves and receive an 'echo' that is reflected back from objects. Radar sensors can provide information about the distance and relative

Figure 4.14 High resolution camera. (Source: First Sensor AG).

speed of these objects. They are very accurate, which makes them ideal for active cruise control (ACC), collision warnings or for emergency brake assist (EBA) systems.

Radar sensors use radio waves so will operate regardless of weather, light or visibility conditions. That makes them an important component in the sensor set. The company ZF, for example, offers a broad

Figure 4.15 Radar sensor on a Tesla. (Source: Bosch Media).

Figure 4.16 Lidar scanning. (Source: ZF TRW).

assortment of sensors with different ranges and opening angles (beam width). This type of sensor is ideal for highly automated and autonomous driving due to its high resolution.

4.3.4 Lidar

Lidar sensors work like radar using the echo principle, except they use laser pulses instead of radio waves. They can record distances and relative speeds just the same as radar. However, objects and angles are recognised with a much higher level of accuracy. The sensors can see complex traffic situations in the dark too. Their angle of view is not critical because they record the 360° environment of the vehicle.

47

Figure 4.17 The Alpha Puck and Velarray. (Source: Velodyne Lidar).

> **Key Fact**
>
> Lidar sensors work like radar using the echo principle.

High-resolution 3D solid state lidar sensors from companies, such as ZF for example, can display pedestrians and smaller objects three dimensionally. This is essential for automation level 4. These sensors use solid-state technology, which is very robust due to the lack of moving components.

Figure 4.17 shows two sensors from the company Velodyne. The first is the Alpha Puck™, a lidar sensor specifically made for autonomous driving and advanced vehicle safety at highway speeds, and the second one is called the Velarray™, which is a powerful lidar for driver assistance.

Key features of the Alpha Puck™ are:

▶ F300m capable sensor for autonomous fleets
▶ Horizontal (360°) and vertical (40°) FOV
▶ Resolution (0.2° × 0.1°) and point density
▶ Proven, Class 1 eye-safe 905 nm technology
▶ Sensor-to-sensor interference mitigation
▶ Dynamic intelligent firing with perception awareness
▶ Bottom connector, with cable length options

4.3.5 Sound

A sensor not used by many manufacturers (yet) is a simple microphone. This is primarily used to recognise emergency vehicle sirens, but it can recognise other sounds. An AI system is required to filter out extraneous noise and to sense the distance and direction. The vehicle can then pull over or move off the road. Moving off road is very complex, however, as the surface structure may be unsuitable.

4.3.6 Ultrasound

Ultrasonic range sensors are often fitted to the vehicle, as shown in Figure 4.18. Generally, a number of intelligent sensors are used to form a detection zone as wide or as long as the vehicle. The ultrasound processing system is relatively simple as the level of discrimination required is low and the sensors only has to operate over short distances.

4.3.7 Combined perception

Cameras, ultrasonic, sound, radar and lidar sensors have advantages and disadvantages. However, if they are combined intelligently, the result is a detailed and reliable 360° view. This prevents blind spots, even in complex situations. When the information from several devices is combined, the accuracy improves and a more complete image with more detail is produced. An ADV must unequivocally comprehend every traffic situation even in poor weather and lighting conditions.

> **Key Fact**
>
> Cameras, ultrasonic, sound, radar and lidar sensors have advantages and disadvantages. However, if they are combined intelligently, the result is a detailed and reliable 360° view.

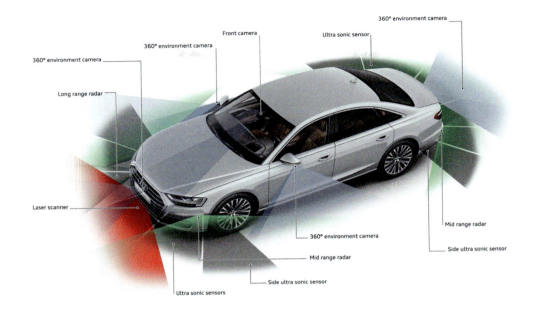

Figure 4.18 Perception areas. (Source: Audi).

Figure 4.19 Mirror replacement blind spot detection camera. (Source: First Sensor AG).

The next section looks at lidar sensors in more detail, and then later there is more detail in section 4.6 on how the combination of sensor signals forms part of the vehicle system architecture.

4.4 Lidar operation

4.4.1 Introduction

Light detection and ranging (lidar) is also known as laser detection and ranging

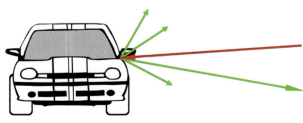

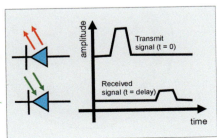

Figure 4.20 Pulsed time of flight (TOF) lidar system

49

(LADAR), time of flight (TOF), laser scanners or laser radar. We will stick with lidar! Just like radar, it is a sensing method that detects objects and their distances. An optical pulse is transmitted, and the reflected return signal is measured (Figure 4.20). The time width of this pulse can vary from a few nanoseconds (ns) to several microseconds (μs).[2]

Definition

Lidar: Light detection and ranging

Light is transmitted from a laser diode in patterns and data is extracted based on the reflections received by the detector. Return pulse power, round-trip time, phase shift and pulse width are common methods used to extract information from the signals.

A spatial resolution on the region of 0.1° is possible with lidar. This is because infrared (IR) light used will collimate as a laser. It also has a short (0.9 to 1.5 mm) wavelength. This means that creating extremely high-resolution 3D images of objects is possible without excessive processing. Radar has a wavelength (4 mm for 77 GHz) so it is difficult to resolve small features, especially at distance.

Key Fact

A spatial resolution in the region of 0.1° is possible with lidar.

Solid-state lidar and radar both have excellent horizontal field of view (FOV), known as the azimuth. Mechanical lidar systems possess the widest FOV of all advanced driver assistance systems (ADAS) technologies because of their 360° rotation. Lidar has a better vertical FOV than radar; this is known as elevation.

Definition

Collimate: To make (rays of light or particles) accurately parallel

As lidar has become more common, prices have dropped considerably from tens of thousands of dollars. It is predicted by some experts that the cost of a lidar module will drop to less than US $200 by 2022.

The mechanical scanning lidar system used a few years ago – commonly seen on Google's self-driving car – was quite bulky. Advances have been made, which have reduced the size, but there is a general shift in the industry towards solid state lidar devices.

Figure 4.21 Photodiode and amplifier. (Source: First Sensor AG).

Figure 4.22 Lidar optical sensor. (Source: First Sensor AG).

Figure 4.23 Google self-driving car with a mechanical lidar sensor mounted on the roof

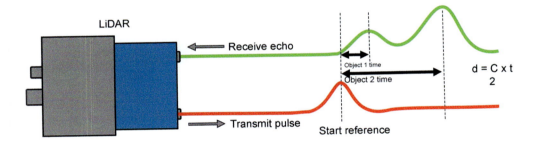

$$d = \frac{C \times t}{2}$$

Figure 4.24 Time of flight (TOF) lidar

Most lidars use direct time of flight measurement. A discrete pulse is emitted, and the time difference between the emitted pulse and the return echo is measured. This can be converted to a distance as shown in Figure 4.24. Optical time of flight measurement is a very reliable option for contact-free measurement of distances.

4.4.2 Types of lidar sensor

There are currently three main types of lidar:

▶ Flash
▶ Micro-electro-mechanical system (MEMS)
▶ Optical phased array

The operation of flash lidar is similar to a digital camera using an optical flash. A single large-area laser pulse lights up the environment in front of it. An array of photodetectors captures the reflected light. The detector is able to determine the image distance, location and reflection intensity. Because this method takes in the entire scene as a single image, the data rate is much faster. Also, as the entire image is

51

captured in a single flash, the method is more immune to vibrations that would distort the picture.

A particular disadvantage of this method is the presence of reflectors on other vehicles. These reflect the majority of the light back rather than some of it scattering, which can overload the sensor. A further disadvantage is the high laser power needed to illuminate the scene to a suitable distance.

The narrow-pulsed time of flight (TOF) lidar method is the most common. Mirrors are used to steer the beams; there are two types in use:

▶ Mechanical lidar, which uses high-grade optics and a moving (rotating) assembly. This creates a wide field of view (FOV), often 360°. The mechanical system results in good signal-to-noise ratio but is bulky.

▶ Solid-state lidar has no moving components. Because of this it has a reduced FOV, so multiple sensors are fitted at the front, rear and sides of a

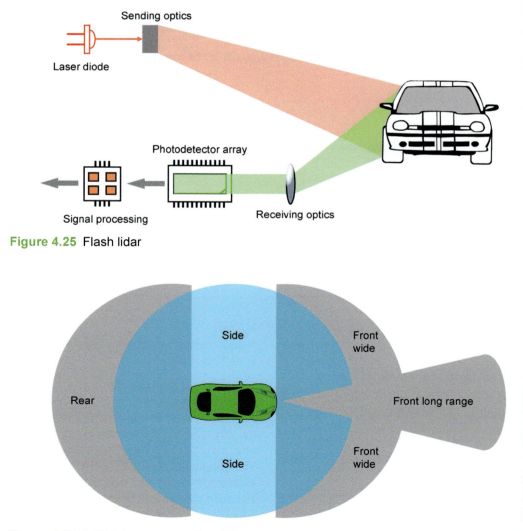

Figure 4.25 Flash lidar

Figure 4.26 360° lidar cocoon using different sensors

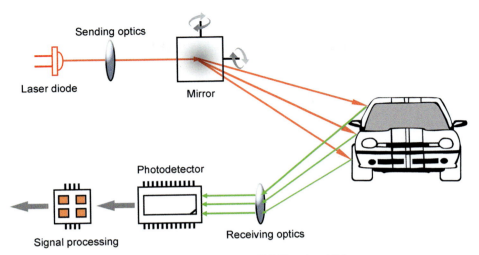

Figure 4.27 Micro-electro-mechanical system (MEMS) pulsed lidar

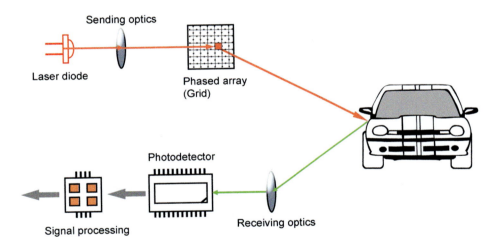

Figure 4.28 Phased array (PA) pulsed lidar

vehicle. The result is an FOV similar to the mechanical system. The FOV is sometimes described as the lidar cocoon.

A micro-electro-mechanical system (MEMS) lidar system uses very small mirrors. The angle of these mirrors can be changed by applying a voltage. The mechanical

scanning hardware is replaced with a solid-state equivalent.

Definition

MEMS: Micro-electro-mechanical system

To move the beam in three dimensions requires several mirrors arranged in a cascade. The alignment process is not simple and can be affected by vibrations. Automotive specifications start at −40°C, which can be a difficult environment for a MEMS device. Developments are ongoing and it is likely to be the sensor of choice for many manufacturers.

4.4.3 Processing system

Figure 4.29 shows a lidar system block diagram. The main subsystems of the lidar signal chain comprise a transmitting system (Tx), a receiving system (Rx) and a custom digital-processing system to extract point-cloud information.

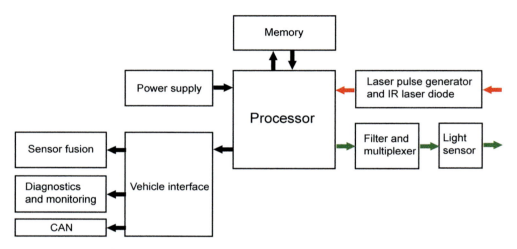

Figure 4.29 Lidar subsystems. (Source: Texas Instruments).

Figure 4.30 Example of a point cloud. (Source: Velodyne).

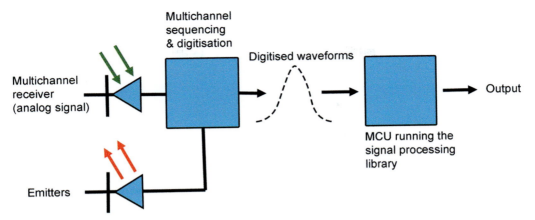

Figure 4.31 Lidar processing system. (Source: LeddarTech).

A point cloud is a set of data points in space. Point clouds are generally produced by 3D scanners, such as lidar, which measure a massive number of points on the surfaces of surrounding objects.

> **Key Fact**
>
> A point cloud is a set of data points in space.

The next sections will examine how data from sensors is processed as part of the automated driving system.

4.5 Sensor positioning

4.5.1 Introduction

The sensors of an ADV determine its operational design domain (ODD), in other words, the different environments in which it is safely used. Determining which sensors are needed for a specific ODD is very difficult because there are so many common cases and edge cases to consider. All sensors have advantages and disadvantages, so any choice is a balance between range, accuracy, FOV, sampling rate, cost and general system complexity.

> **Definition**
>
> ODD: Operational design domain

4.5.2 Scenarios

The best way to work out what sensors are needed for an ODD is to capture a variety of scenarios.[3] These are then used as data for a simulation to find out the circumstances where a sensor set would miss something. The performance of different layouts of sensors can be compared, and the best ones are gradually improved. This is known as a data-driven approach and can use machine learning in order to identify the best configurations. Scenarios can be designed manually or collected during field trials. A selection of difficult scenarios and edge cases (ODDs) will now be examined briefly.

Obstacles in a vehicle's path that are not connected to the ground, for example, open doors, protruding freight, tree branches, barrier gates or low bridges, are difficult. They can be missed completely by low mounted or low resolution lidars. They are very difficult for cameras and can even be a challenge for some radars.

55

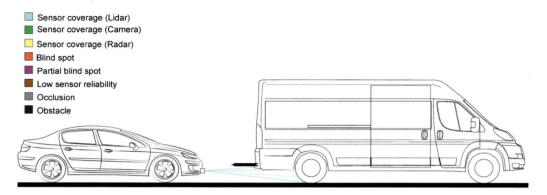

Sensor coverage (Lidar)
Sensor coverage (Camera)
Sensor coverage (Radar)
Blind spot
Partial blind spot
Low sensor reliability
Occlusion
Obstacle

Figure 4.32 If a lidar sensor with small vertical FOV is mounted too low the obstacle is not sensed

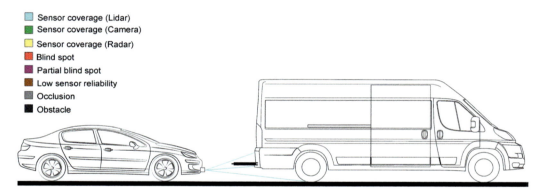

Sensor coverage (Lidar)
Sensor coverage (Camera)
Sensor coverage (Radar)
Blind spot
Partial blind spot
Low sensor reliability
Occlusion
Obstacle

Figure 4.33 If the vertical resolution of a lidar sensor is too low the obstacle is not sensed

Sensor coverage (Lidar)
Sensor coverage (Camera)
Sensor coverage (Radar)
Blind spot
Partial blind spot
Low sensor reliability
Occlusion
Obstacle

Figure 4.34 Cyclists riding next to the vehicle could be missed by a roof mounted lidar sensor

When turning left or right, extra caution must be taken because of traffic that could be cut up. Large vehicles are quite easy to detect, but cyclists, for example, are difficult to see using roof centre sensors.

To reduce the risk of collisions to a when entering an occluded crossing, it is important to be able to observe cross-traffic lanes as soon as possible. Roof sensors may miss these so sensors mounted at

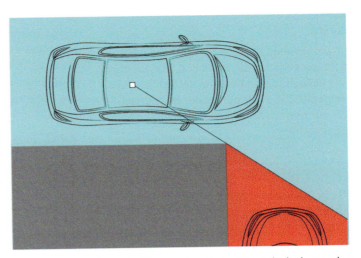

Figure 4.35 A centrally mounted sensor cannot see traffic coming from an occluded crossing lane

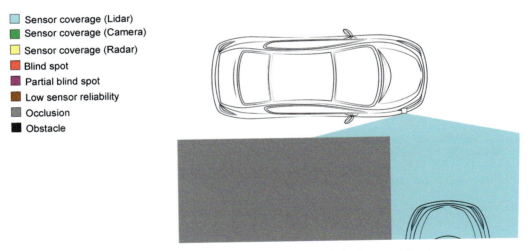

Figure 4.36 A sideways facing sensor at the front can see into the crossing lane

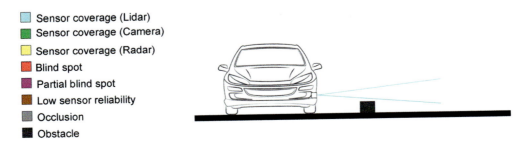

Figure 4.37 A low obstacle is missed by the sensor

the front of the vehicle are a more sensible choice.

Low obstacles that are close to the car, such as kerbstones, or blocks in a car park are difficult to see by sensors with a small vertical FOV. They can either be missed, or perhaps worse, they can occlude traffic behind them.

Key Fact

Low obstacles that are close to the car are difficult to see by sensors with a small vertical FOV.

The slope of the road can have a serious effect on the vertical FOV of sensors. When approaching ramps, tunnels or car

parks for example, the coverage of other traffic is limited. Also, ramps could be classified as obstacles if the road surface is not compensated for. Particularly difficult situations are intersections with lanes that are ascending or descending.

Definition

FOV: The field of view is the extent of the observable world that is seen at any given moment. In the case of optical instruments or sensors it is a solid angle through which a detector is sensitive to electromagnetic radiation.

Sensor coverage (Lidar)
Sensor coverage (Camera)
Sensor coverage (Radar)
Blind spot
Partial blind spot
Low sensor reliability
Occlusion
Obstacle

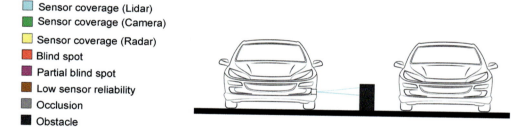

Figure 4.38 A low obstacle occludes traffic behind it

Sensor coverage (Lidar)
Sensor coverage (Camera)
Sensor coverage (Radar)
Blind spot
Partial blind spot
Low sensor reliability
Occlusion
Obstacle

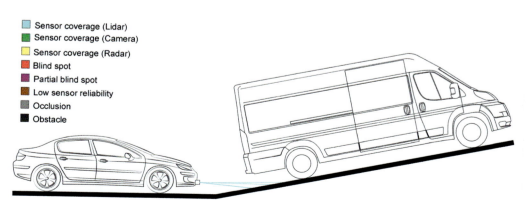

Figure 4.39 A ramp in front of a vehicle can be interpreted as an obstacle, and the vehicle is missed by the sensor

4.5.3 Lidar

Introduction

Positioning a lidar sensor determines its FOV. There are four main ways in which this sensor can be used. Each of which will be examines briefly.

Single lidar on roof centre

In early ADVs this seemed to be the method of choice, but it is not very effective unless the car shape is changed, because it has blind spots.

A sensor can be moved higher to reduce the blind spot caused by the roof. The sensor can also be moved from the centre to the front. This improves front perception but decreases rear.

This method is simple to set up, and makes it relatively easy to synchronise and align multiple point clouds. It gives 360° coverage with a single sensor and gives a good overview of other traffic.

However, it has blind spots for low objects in all directions, particularly to the rear if fitted at the front. This type of sensor must therefore be raised above the roof to get full vertical FOV. This doesn't look very attractive, has mechanical challenges and puts low car parks out of bounds!

Multiple lidars on the roof

Using multiple roof sensors can mitigate some of the disadvantages of the single centre mounted method.

Sensors fitted in this way can be tilted to decrease blind spots. They can be mounted on the edges of the roof. The occlusion by the vehicle roof is eliminated, and if tilted, they can reach optimal coverage for a roof mounted system. However, there is more complexity in terms of integration and point cloud fusion compared to a single sensor.

Front lidar

Most manufacturers now use front mounted lidar as part of their sensor package.

Lidars with a larger vertical FOV enable steeper slopes to be detected correctly. If mounted higher, such as above the windscreen at the front of the roof, then it can look over traffic and low obstacles. This method is quite (relatively) simple

□ Sensor coverage (Lidar)
□ Sensor coverage (Camera)
□ Sensor coverage (Radar)
■ Blind spot
■ Partial blind spot
■ Low sensor reliability
■ Occlusion
■ Obstacle

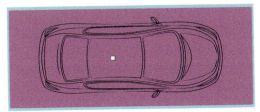

Figure 4.40 Roof centre lidar

Figure 4.41 Multiple roof lidars

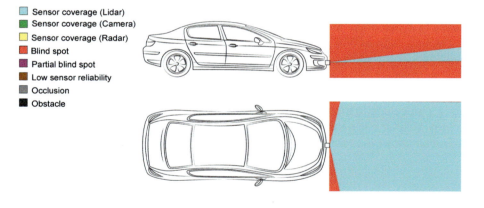

Figure 4.42 Front lidar

to set up and integrates well into the vehicle. Lidar beams parallel to the road surface allow the detection of obstacles and measurement of the distance to them. On slopes the sensor could sense the ground plane as an obstacle for ascending roads, and it gives no returns for descending roads. It is not adequate for urban use unless combined with other sensors.

Side view lidars

These lidars are useful for detecting objects such as pedestrians or cyclists.

If a higher FOV for both horizontal and vertical is used, this gives good coverage of the vehicle's sides. Integration in mirrors, for example, allows it to see over low obstacles. The vehicle can detect cross-traffic at

Sensor coverage (Lidar)
Sensor coverage (Camera)
Sensor coverage (Radar)
Blind spot
Partial blind spot
Low sensor reliability
Occlusion
Obstacle

Figure 4.43 Side view lidars

intersections (see above scenario) which is a useful feature, but sensors with low vertical FOV will have issues with slopes at the vehicle sides.

4.5.4 Camera

Introduction

Cameras (lidars, radars and ultrasonic sensors are active) are passive sensors. Just like our eyes, they collect light reflected from the environment. Lack of light (darkness), rain, fog, low sun, spray, snow, dust, insects and more can influence the operation of a camera. However, they can offer a high resolution.

> **Definition**
> CMOS: Complementary metal –oxide–semiconductor is a technology for constructing integrated circuits. It can also be used for analog circuits such as image sensors.

Camera images need powerful processors to interpret what they see. Deep learning is often used, and this is covered in section 4.10. Data from distance measuring sensors such as radar lidar can be used with much less processing for collision avoidance. However, camera data provides the best representation of the vehicle's environment.

> **Key Fact**
> Camera images need powerful processors to interpret what they see.

Figure 4.44 High resolution camera that uses CMOS technology. (Source: First Sensor AG).

Wide-angle lenses can be used, and these have a large FOV. Unfortunately, these

61

images cause distortion which has to be corrected before full processing. Also, areas close to the edges of the image may be blurred and less reliable for image recognition tasks.

Roof mounted

Some production cars are equipped with a set of wide-angle cameras to provide a bird's-eye view of the close environment to make parking easier.

Several 180° cameras are mounted on the vehicle, giving a 360° horizontal coverage.

For ADAS, cameras are mounted around the body of the vehicle, while for self-driving they are more likely to be mounted on the vehicle's roof. These cameras can be tilted to cover the close environment. There are options to mount them on the corners of the vehicle or the roof edge centres.

The main advantages of this method are the 360° coverage and the top down view to support parking. However, 180° cameras are subject to lens distortion; resolution per degree is reduced and this can limit the detection range. Cameras at the roof corners can have larger areas of low sensor

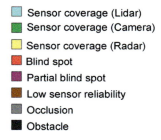
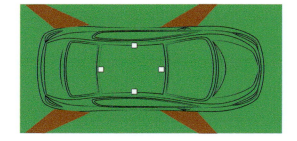

Sensor coverage (Lidar)
Sensor coverage (Camera)
Sensor coverage (Radar)
Blind spot
Partial blind spot
Low sensor reliability
Occlusion
Obstacle

Figure 4.45 Roof cameras

Figure 4.46 Four cameras connected to a control unit. (Source: First Sensor AG).

reliability compared to cameras on the roof edge centres.

Front mounted

Front cameras are the main sensor for environment recognition in current production cars. ADAS systems such as active cruise control and automated dipping headlights can also use information from front cameras.

Front mounted cameras are relatively cheap, but like all cameras, they are limited by environment conditions. Front cameras are typically mounted between the rear mirror and the windscreen; they are often part of the mirror assembly. The screen protects the camera and is cleaned by the washers and wipers. Front cameras can also be installed inside the vehicle between the dashboard and the screen, outside on the bumper or at the centre of the front roof edge. Stereo cameras are usually used because they provide distance estimation.

Key Fact

Front cameras are typically mounted between the rear mirror and the windscreen.

Front cameras benefit from the vehicle's headlights at night. A range of ADAS features like lane departure warning, lane change assistant and lighting adjustment are facilitated. If they are behind the screen, which is the most common option, they are protected from rain and dirt. However, their vertical FOV is limited because of the bonnet, and small objects in front of the vehicle can be occluded. Like all sensors they need to be combined with others for urban automated driving.

Traffic lights

Front camera and surround view camera systems have a limited vertical FOV so are usually unable to detect objects such as traffic lights.

A wide-angle camera mounted at the front roof edge centre, if tilted upwards as necessary for the environment, can achieve traffic light detection.

4.5.5 Radar
Introduction

Radar is already an established technology in the automotive industry. It has been used for many years to enable ADAS features

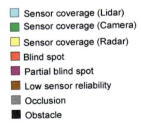

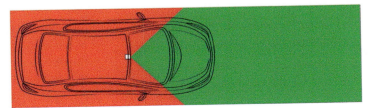

- ☐ Sensor coverage (Lidar)
- ☐ Sensor coverage (Camera)
- ☐ Sensor coverage (Radar)
- ☐ Blind spot
- ☐ Partial blind spot
- ☐ Low sensor reliability
- ☐ Occlusion
- ☐ Obstacle

Figure 4.47 Front mounted camera sensor coverage

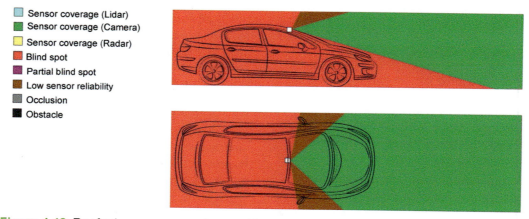

Figure 4.48 Roof edge camera can be used for traffic light detection

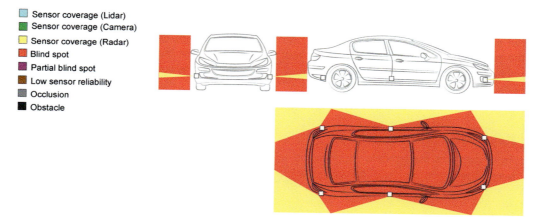

Figure 4.49 Radar coverage using short- and mid-range

such as adaptive cruise control (ACC) and autonomous emergency braking (AEB). They accurately measure distance and velocity. They are particularly good at detecting metallic objects like vehicles but can detect non-metallic objects such as pedestrians with a reduced range.

Definition

AEB: Autonomous emergency braking

Short- and mid-range radars

It is possible to achieve 360° coverage by using short- and mid-range radars.

Four to six short range radars (SRR has about a 30 m range) and mid-range radars (MRR has about a 100 m range) can achieve a 360° horizontal coverage. In particular, this can be very good in many typical urban ODDs. However, the narrow vertical FOV results in difficulties when in non-flat terrain.

Front radar

Mid- and long range radar can be combined at the front of the vehicle.

Mid-range and long range radars enable ADAS features like ACC and AEB. Radar is better than lidar and cameras in environment conditions like rain, fog and snow. Additional sensors are needed for urban ODD.

4.5.6 Summary

It will have become clear by now that there is not a perfect sensor configuration for all situations and budgets. There are multiple configurations that work, but of course all have advantages and disadvantages. Arguably this is the case for all technological developments: there is no perfect solution, just a better one each time.

4.6 Automated driving system

Like all complex systems, it is helpful to consider an automated driving system as a block diagram. One single block could be used to represent the whole vehicle, but using the standard input, control, output method is more useful.

The sense block, in Figure 4.51, represents the combined range of sensors used by the vehicle. These will be a combination of some or all of those discussed in the previous sections, and in a range of positions.

The understand block is where the serious processing takes place. Significant digital processing is needed because as well as the obvious need for safety, the quantity of data collected by the sensors is huge. As mentioned earlier, a driverless vehicle

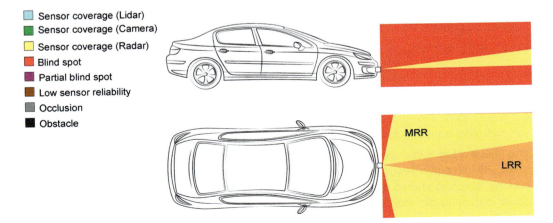

Legend:
- Sensor coverage (Lidar)
- Sensor coverage (Camera)
- Sensor coverage (Radar)
- Blind spot
- Partial blind spot
- Low sensor reliability
- Occlusion
- Obstacle

Figure 4.50 Combining two front radars

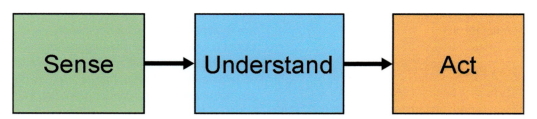

Figure 4.51 The standard block diagram for an ADV: Inputs – Control – Outputs

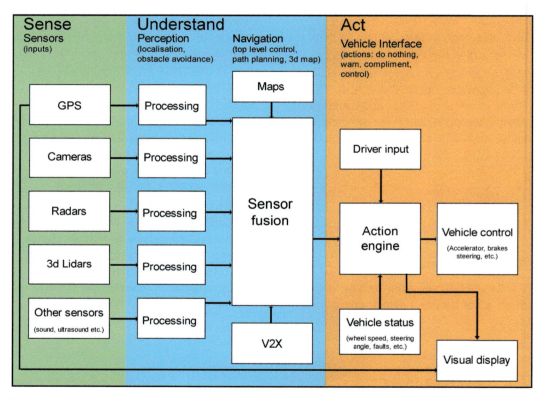

Figure 4.52 ADV system block diagram

travelling down a high street for a mile generates more data that an aeroplane flying (on autopilot) from London to New York!

The final act block incorporates all aspects of the vehicle controls such as its speed, braking, steering and so on. It also covers a driver display and/or warning devices

Working on the same block diagram principle it is now possible to look in

more detail at what is in each of the three main blocks. This is shown as Figure 4.52.

The Sense block in Figure 4.52 is a range of sensors, which in this case are:

▶ GPS receiver to give navigation positional and a compass for direction of travel
▶ Cameras to see the general surroundings (stereo cameras perceive depth)

▶ Radar to build a picture of the longer range surroundings even in poor light or bad weather
▶ Lidar to build up a 3D image of the surroundings
▶ Other sensors as appropriate to the particular design.

Information from these sensors is processed individually and then combined or fused in the Understand block. Maps come from stored data just like on a normal navigation system, except the need for regular updates is very important. V2X is shorthand for vehicle to everything. Updates can be downloaded from the cloud, and information from other connected cars can be processed.

Key Fact

GPS maps come from stored data just like on a normal navigation system, except higher definition is often needed for ADVs.

The Act block on the right hand side of the diagram is where the vehicle is finally controlled – or driven. The action engine

combines information from sensor fusion with driver input and live vehicle status such as speed or steering angle. It then outputs to a visual display and the actual vehicle control actuators, the key outputs being the:

▶ Accelerator (on an ICE car this is a throttle actuator; more likely it is a signal to the electric vehicle ECU)
▶ Brakes (the ABS pump and control valves)
▶ Steering (the EPAS motor).

The level of driver input will depend on the level of automation.

4.7 Mapping

A key issue is not just what the car sees, but what it knows beforehand about the area it is travelling through. High-definition 3D maps are, therefore, an essential element for the navigation and safety of these vehicles. Accuracy down to a centimetre is needed.

An ADV driving through a high street collects more than a terabyte of data a day, which is the equivalent of about 1,400 CDs

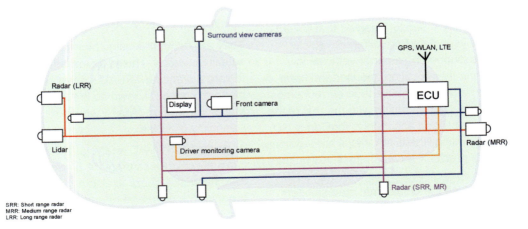

Figure 4.53 A range of sensors used on the car results in more accuracy and the elimination of blind spots

full. It is not practical to send this quantity of data across the internet, so it is generally physically moved from one hard drive to another. This method is sometimes referred to as the 'sneakernet' because it moves at the pace of the engineer's footwear!

Storage of this quantity of data is just one of the technical issues challenging engineers. These maps are so important for ADVs not just for location, but because it reduces the amount of work that the vehicle software has to do to recognise the world. By comparing actual surroundings to what is in the map, they can focus more attention on things that are different, like identifying a pedestrian, an animal or a bike.

Creating high resolution maps can be done in a number of ways; here we will examine two different approaches taken by some leading companies:

▶ Crowdsourcing
▶ Satellites.

TomTom[4] and Qualcomm[5] recently announced a project to crowdsource high-definition map data. TomTom's HD Map for autonomous vehicles will benefit from a richer set of data due to Qualcomm Technologies offering a new chip to be used in car sensors.

Qualcomm Technologies' Drive Data is a platform for car sensor analysis that collects and analyses data from different vehicle sensors, supporting smarter vehicles to determine their location, monitor and learn driving patterns, perceive their surroundings and share this perception with the rest of the world reliably and accurately. TomTom's HD Map, including RoadDNA, helps autonomous vehicles precisely locate themselves on the road and plan manoeuvres. TomTom's HD Map can use the Qualcomm map snippets for maintenance.

Traditional development of maps requires deploying dedicated fleets of vehicles that are equipped with professional-grade sensors to collect location, raw imagery, lidar and other data, which is then transferred, stored and processed in data centres. Now that cars are increasingly connected and equipped with a range of sensors, new and complimentary approaches become possible.

Using the capabilities of the Qualcomm Drive Data Platform, which features the Qualcomm® Snapdragon™ 820Am automotive processor, TomTom and Qualcomm Technologies aim to facilitate adding an improved, scalable and cost-efficient crowdsourcing approach to the mix of sources for HD map making.

The new concept is designed to allow massive numbers of connected cars to see and understand their environment, traffic and road conditions, and support real-time input for map and road condition updates.

A group of companies, Toyota Research Institute-Advanced Development (TRI-AD), Toyota's automated driving software development company, Maxar Technologies (Maxar), a global technology innovator powering the new space economy, and NTT DATA Corporation (NTT DATA), a leading IT services provider, are collaborating on a proof of concept building automated high-definition maps for ADVs.

They are doing this using high-resolution satellite imagery. This is an important move toward advancing TRI-AD's open software platform concept known as the Automated Mapping Platform (AMP), which will help the scalability of ADVs.

According to TRI-AD analysis, HD maps currently cover less than 1% of the global road network.[6] There is also a need to broaden the coverage of urban areas and local roads before ADVs can become mainstream. An HD map created from the accurate satellite imagery allows the driving software to compare multiple data sources

Figure 4.54 Enhanced 3D map. (Source: TomTom).

Figure 4.55 Example of Tokyo region satellite image. (Source: Toyota).

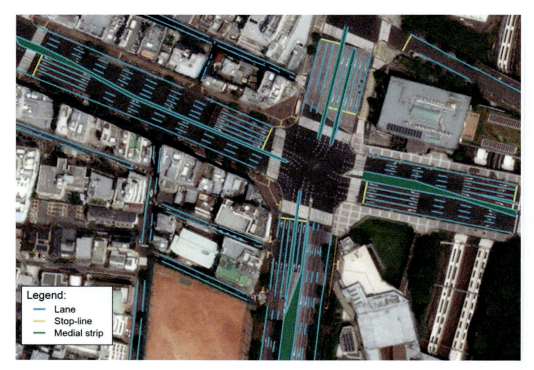

Legend:
— Lane
— Stop-line
— Medial strip

Figure 4.56 Example of high definition map for automated driving. (Source: Toyota).

and signal the car to act in such a way as to stay safe.

In this proof of concept, the three companies will work together to process satellite imagery into vehicle-friendly HD maps. Using Maxar's cloud-based Geospatial Big Data platform (GBDX), imagery from Maxar's optical satellite imagery library will be fed into NTT DATA's specialised algorithms. This AI will extract information that is necessary to generate the detailed road network. TRI-AD will make HD maps available for delivery from their cloud into Toyota test vehicles.

The project started with the creation of an HD map for a pre-defined area of the Tokyo metropolitan region (Figures 4.55 and 4.56). This then opens the possibility of supporting automated mobility on all roads around the world.

Recent advances in electronics and aerospace engineering are leading to higher resolutions and more frequent updates of global imagery from space-based assets. Additionally, machine learning is helping automate the discovery and integration of semantic relationships between road elements within image data.

4.8 Other technologies

4.8.1 Platooning

An EU project called SARTRE was launched to develop and test technology for vehicles that can drive themselves in long road trains on motorways. This technology has the potential to improve traffic flow and journey times, offer greater comfort to drivers, reduce accidents and improve fuel consumption and hence lower CO_2 emissions.

Definition

SARTRE: Safe Road Trains for the Environment

This technology, known as platooning, is one of the many steps towards full autonomous driving. Figure 4.57 illustrates why automated vehicles can travel closer together.

Definition

V2V: Vehicle to Vehicle

The automotive industry has long been focused on the development of active safety systems that operate preventively, traction control and braking assistance programs, for example. Platooning technology means that the vehicle is able to take control over acceleration, braking and steering, and can be used as part of a road train of similarly controlled vehicles.

The vehicles are equipped with a navigation system and a transmitter/receiver unit that communicates with a lead vehicle. Since the system is built into the cars, there is no

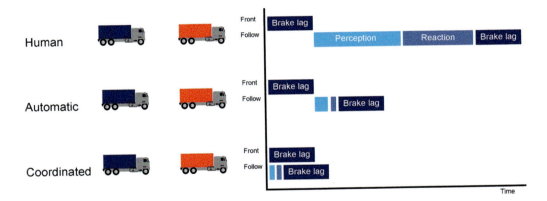

Figure 4.57 V2V communication allows close platooning

Figure 4.58 One car following the lead vehicle. (Source: Volvo Media).

need for any infrastructure change to the road network.

The idea is that each road train or platoon will have a lead vehicle that drives exactly as normal, with full control of all the various functions. This lead vehicle is driven by an experienced driver who is thoroughly familiar with the route. For instance, the lead may be taken by a taxi, a bus or a truck. Each such platoon or road train will consist of six to eight vehicles.

As a driver approaches their destination, they take over control of the vehicle and leave the convoy by exiting off to the side. The other vehicles in the road train close the gap and continue on their way until the convoy splits up.

The advantage of these road trains is that all the other drivers in the convoy have time to get on with other business while on the road, for instance when driving to or from

work. The road trains increase safety and reduce environmental impact thanks to lower fuel consumption compared with cars being driven individually. The reason for this is that the cars in the train are close to each other, exploiting the resultant lower air drag. The energy saving is expected to be in the region of 20%. Road capacity will also be able to be utilised more efficiently.

As the participants meet, each vehicle's navigation system is used to join the convoy, where the automated driving program then takes over. As the road train approaches its final destination, the various participants can each disconnect from the convoy and continue to drive as usual to their individual destinations. Figure 4.59 outlines the process of joining and leaving a platoon.

The tests carried out (2011) included a lead vehicle and single following car. The

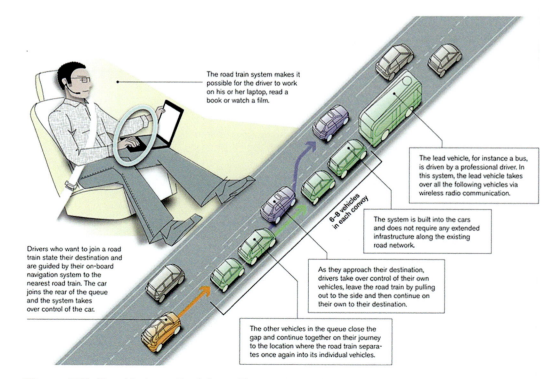

Figure 4.59 Road train methodology. (Source: Volvo Media).

steering wheel of the following car moves by itself as the vehicle smoothly follows the lead truck around a test track. The driver is able to drink coffee or read a paper, as no input is required to operate the vehicle.

The platooning technique is designed to achieve a number of things:

- Road safety, as it minimises the human factor that is the cause of at least 80% of accidents
- Fuel consumption and CO_2 emissions improve by up to 20%
- Convenience for the driver as it frees up time for other activities
- Traffic congestion will be reduced because the vehicles travel at highway speed with only a few meters gap.

4.8.2 Geofencing

A possible development for the near future is that ADVs will be geofenced such that they will only operate at certain automated driving levels in specified areas. This could be on some stretches of motorway, for example, but not in town centres.

Geofencing is a system where an application or other software program uses GPS to trigger a pre-programmed action when a vehicle enters or leaves a specified area. For example, the system may disable or enable some automated driving features.

4.8.3 Passive safety

Passive safety systems are those that only activate in the event of an accident or potential accident. A typical passive safety example is the airbag. On normal vehicles positioning the airbags is fine because the layout and structure is fixed. However, with some high driving-level vehicles it is likely that the steering wheel could be retracted, and seats may be allowed to pivot so that passengers can sit facing each other.

Key Fact

Passive safety systems are those that only activate in the event of an accident or potential accident.

This new layout will require different methods of passenger protection. For example, airbags would not be located inside the steering wheel but installed all across the front or all around the inside of the vehicle. Seat belts will need to have different fixings.

Future airbags will need to be more compact and lighter as the electronic components in cars will take up more room. They may even need to inflate and deflate, in other words be reusable.

Manufacturers of car safety systems have already started working on designing new solutions to meet and anticipate the needs of this changing market.

4.8.4 Potholes and debris

The amazing sensor technology used on ADVs will arguably miss one important area – the road surface. A human driver can see the road surface in enough detail to take evasive action, if safe to do so, around a deep pothole or piece of unknown debris. Perhaps this omission is because a lot of development takes place in areas with nice roads!

It is possible to fit sensors that would 'look' at the road, but this would add extra expense and a whole new level of processing needs. One interesting solution was developed some time ago by Ford engineers. Accelerometers are fitted on the suspension and can detect that a wheel is dropping into a pothole within just a few milliseconds. It then changes the damping rate such that the wheel does not drop all the way into the hole and consequently hits

Figure 4.60 Road surface

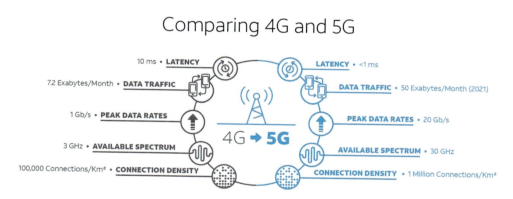

Figure 4.61 Comparing 4 and 5G. (Source: https://policyforum.att.com).

the far edge of the hole with less force and at a shallower angle. This reduces the risk of damage.

4.9 Connectivity

4.9.1 5G networks

5G follows previous mobile network generations 2G, 3G and 4G. Compared to current networks (mostly 4G and 3G technology), 5G is designed to be faster and more reliable. It also has greater capacity and lower response times.[7]

> **Definition**
>
> 5G: Fifth generation mobile data network

The key benefit of 5G is speed, which can be in excess of 1 Gb/s (1000 Mbit/s) and could reach ten times this figure. The speed we experience in the real world depends on many factors, such as how far we are from a base station and how many other people are using the network at the same time. However, a typical user experienced data rate for downloads is expected to be a minimum of 100 Mbit/s (which is much faster than existing systems).

There is potentially an even bigger benefit with 5G and that is low latency. 4G and other network standards have a long delay between a command being issued and a response being received. This is about 65 ms for 3G and 40 ms for advanced 4G. Fixed 'landline' broadband connections in the UK have a latency of 10 ms to 20 ms. 4G has a target of 4 ms, but for mission critical applications (like ADVs) this can be 1 ms.

5G will also have access to more spectrum and at higher frequencies, in particular millimetre wave, which is the band of between 30 GHz and 300 GHz. This means that networks will be able to handle many high-demand applications all at once.

Table 4.1 Generations compared

Network	Max download speeds	Download an HD film
3G	384 Kbps	More than a day
4G	100 Mbps	7 minutes+
4G+	300 Mbps	2.5 minutes
5G	1–10 Gbps (theoretical)	4 to 40 seconds

Figure 4.62 Vehicle occupants will exchange lots of different types of data so security of the link will be essential. (Source: IBM).

Because of these improved speeds and short latency automakers will be able to download more data and use the cellular network for some safety-related vehicle-to-everything (V2X) features.

Interestingly, data going the other way, from passengers to automakers, could be a valuable commodity. They may also make more money on services over time, like selling films or other entertainment. Cellular V2X is likely to start by using 4G links, but it is the transition to 5G's higher bandwidth capabilities that will facilitate the extensive communications needed for autonomy and other features and functions.

Features like tele-operated driving and cooperative manoeuvres will become possible. Data collected from vehicle cameras, meanwhile, can be collated to create maps that include real-time updates such as accidents and temporary roadworks. Because of the higher bandwidth, large high resolution map tiles could be uploaded and downloaded as needed.

Cybersecurity will be a major issue, and most experts believe communications will have to be stored using techniques such as blockchain. This will ensure that information in messages cannot be tampered with, and that IDs are trusted.

4.9.2 Navigation NDS data standard

The map display on built-in navigation systems is becoming even more engaging and relevant. Buildings extend skyward, enabling you to get your bearings more easily, and visible changes in terrain height combined with integrated satellite imagery

Figure 4.63 V2X can prevent accidents. (Source: Continental).

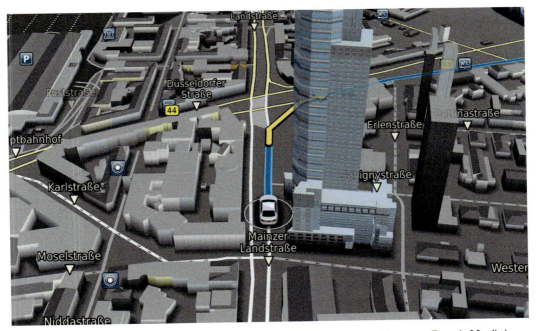

Figure 4.64 Powerful 3D map engine supports continuous zoom. (Source: Bosch Media).

produce an almost photorealistic look. This is made possible by Bosch's advanced navigation software, which takes data compliant with the new Navigation Data Standard (NDS) and processes it in a 3D rendering module to turn it into a visually stunning map.

Definition

NDS: Navigation Data Standard

The system will work offline but if an internet connection is available, the system can enhance the map display with dynamic data. In the future, this will allow integration of the latest weather information or prices at fuel stations along the route, for example.

The key component in the new navigation software is a 3D map engine based on OpenSceneGraph. It superimposes three-dimensional elements like buildings using an additional display layer and can also

make them transparent, keeping the route visible to the driver when it goes behind structures. The driver can smoothly zoom the visible map area, from the highest level of detail to the world view. Using topographical information contained in the NDS data, the software displays differences in terrain height. It will even be possible to artificially bend up the map in the direction of the horizon, thus maximising the amount of screen area used to display the route.

Key Fact

The key component in the new navigation software is a 3D map engine based on OpenSceneGraph.

For interacting with the system, the driver can choose between voice input, multi-touch and handwriting recognition. It is also possible to show different areas of the map on different screens at the same time, such

77

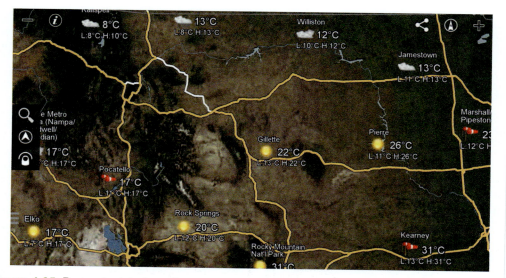

Figure 4.65 Dynamic data from the connected horizon – more than just traffic info. (Source: Bosch Media).

as the displays in the centre console and instrument cluster. The level of display detail can be adapted to the infotainment system's computing power and memory. The navigation software can thus be configured to suit carmakers' requirements.

Traffic congestion can already be portrayed on the map in near real time. But if the infotainment system has internet access, it will be possible in the future to integrate even more information in the map display. The Bosch connected horizon, for example, gives real-time access to data on road conditions stored in the cloud. The 3D map engine can visualise this data, so that areas of the map appear in a different colour if there is particularly heavy rain or a risk of black ice.

In electric vehicles, the system uses a coloured, transparent overlay to indicate the current range on the map for the remaining battery charge.

4.9.3 Vehicle to vehicle (V2V)

Connected vehicles can exchange data via wireless communication. The system known

as dedicated short range communication (DSRC) offers the most reliable technology for this vehicle-to-vehicle (V2V) communication for safety. DSRC transmits messages via a type of Wi-Fi that is similar to office or home systems. The messages can be transmitted to other vehicles over a longer range, and through barriers such as heavy snow or fog. DSRC, as well as other forms of connected vehicle communications, also make it possible to link vehicles with infrastructure. This enables coordination and cooperation that can reduce congestion and improve traffic flow. Pedestrians and bicyclists can be linked in through portable devices.

Definition

DSRC: Dedicated short range communication

The car of the future will be connected. This is because using up-to-the-minute information from the internet will get vehicle occupants to their destination

Figure 4.66 Bosch technology puts the car online. (Source: Bosch Media).

even more safely, efficiently and conveniently. Integration into the internet of things also unlocks a host of vehicle-related services.

Bosch have a cloud-based alert that warns drivers within ten seconds if there is a wrong-way driver approaching. The warning system is a connected lifesaver in the true sense of the word.

To connect the car with the internet, Bosch pursues two main approaches. First, it makes full use of the driver's smartphone. Using the integrated mySPIN solution, drivers can link their Android and iOS devices to the vehicle's infotainment system. Selected apps can then be conveniently operated from the vehicle's central display. This technology has been featured in Jaguar and Land Rover models since 2014.

Bosch's second approach constitutes equipping the vehicle with connectivity hardware in the form of a connectivity control unit, or CCU. The CCU receives

and transmits information using a wireless module equipped with a SIM card. It can also determine the vehicle's position using GPS if desired.

> **Definition**
>
> CCU: Connectivity control unit

Connected to the vehicle's electrical system via the OBD interface, the CCU is available both as original equipment and as a retrofit solution. This makes it possible for fleet operators to retrofit their existing vehicles as well. The Bosch subsidiary Mobility Media also markets this solution for private users under the name Drivelog Connect. A smartphone connected to the CCU can display vehicle data, offer tips on fuel-efficient driving and, in the event of a breakdown,

79

Figure 4.67 A connected car drives more proactively than a human!. (Source: Bosch Media).

immediately contact a towing service and the garage if required.

Information on traffic jams, black ice and wrong-way drivers is available in the cloud. When combined with infrastructure data from parking garages and charge spots, this provides a broader perspective known as the connected horizon. In the connected vehicle, the driver can 'see' over the top of the next hill, around the next bend and beyond. Because future cars will warn drivers in plenty of time about sudden fog or about a line of cars stopped behind the next bend, driving will be safer.

Connectivity also enhances vehicle efficiency. For example, precise data about traffic jams and the road ahead makes it possible to optimise charging management in hybrid and electric vehicles along the selected route. And because the car thinks ahead, the diesel particulate filter can be regenerated just before the car exits the motorway, and not in the subsequent stop-and-go traffic. Connectivity improves convenience as well, as it is a prerequisite for automated driving. It is the only way to provide unhurried braking in advance

of construction zones, traffic jams and accident scenes.

Key Fact

Connectivity enhances vehicle efficiency.

Along with driving data and information on the vehicle's surroundings, the connected car also captures data on the operation of individual components. Running this data through sophisticated algorithms permits preventive diagnostics. For example, the data collected from an injection nozzle can be put through distributed algorithms in the cloud and in the vehicle to predict the part's remaining service life. The driver or fleet operator can be notified immediately, and an appointment made with the workshop in good time. In this way, it is often possible to avoid expensive repair and down times, especially for large commercial vehicles.

Buildings, hedges or a truck can quickly obscure drivers' view, especially at intersections. If a road user is driving carelessly, it is often a matter of milliseconds that determine whether there is a collision or not. However, vehicle connectivity can greatly reduce the number of resulting accidents by promptly providing information that is outside the driver's and the vehicle's field of vision.

Figure 4.68 Cloud connection. (Source: Bosch Media).

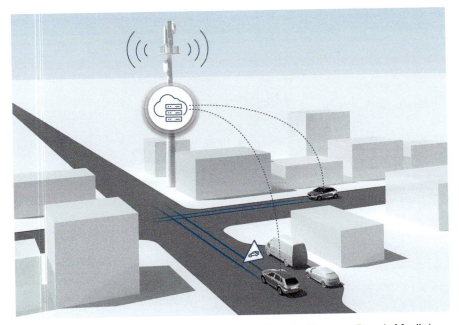

Figure 4.69 Local 'cell based' cloud intersection assistant. (Source: Bosch Media).

Together with Nokia and Deutsche Telekom, Bosch is developing local cloud solutions for the automotive industry and working on the complete integration of vehicles via the cellular network all the way through to the Bosch IoT Cloud. The companies are employing Mobile Edge Computing (MEC), a cellular network technology that uses a local cloud to aggregate and process latency-critical information and distribute it to drivers. Unlike most clouds, this local cloud is situated directly at a mobile base station near the roadside and not on the internet.

Vehicles must able to communicate with each other and via a server, in either a central or a local cloud. For the intersection assistant to work, vehicles must regularly send their location and movement data to the server. This data is compared with that of nearby vehicles, as well as the rules governing right of way. If there is danger of an accident occurring, a warning message is displayed in the vehicle that does not have the right of way.

Outside of cities, where vehicles travel at higher speeds, there is a definite advantage if data takes the short route via the local cloud. Compared to solutions that exchange information via a central cloud, local cloud approaches are at least three times faster, and they have much lower variances in the case of vehicle-to-vehicle latencies under 2 ms. In some situations, this can make the difference as to whether the information reaches the car on time and the driver or the safety function can react quickly enough.

Key Fact

High vehicle-to-vehicle latencies can prevent the system reacting quickly enough.

4.9.4 Vehicle-to-everything (V2X)

To enable connected and automated driving in the future, vehicles must be able to easily communicate with one another as well as

Figure 4.70 Communication with everything. (Source: Bosch Media).

with their surroundings. There is currently no globally standardised technical basis for this exchange of data, which is known as vehicle-to-everything communication, or V2X. Instead, vehicles will in future communicate using the wide variety of different standards implemented by countries and vehicle manufacturers around the world.

However, Bosch has combined connectivity units and telematics units, which individually are only capable of a single transmission technology, to create an all-in-one central control unit for V2X data communication. Cars can then use the Wi-Fi networks available in cities, while elsewhere they can communicate using, for instance, cellular networks.

The complex task of managing these diverse communication options is handled by a software solution from the Silicon Valley-based start-up Veniam. It continuously searches for the best

transmission technology that suits the particular requirements and switches automatically between the available alternatives. The software therefore maintains continuous and seamless vehicle connectivity, ensuring cars can, for example, reliably alert one another to accidents and passengers can enjoy uninterrupted music streaming.

It is expected that the number of connected vehicles on the roads in Europe, the United States and China alone will exceed 470 million by 2025.[8] Initially, most vehicles will connect directly to the cloud. In the near future, increasing numbers of vehicles will also be able to communicate directly with one another as well as with traffic signals, road construction sites, pedestrian crossings and buildings, etc. They will then be able to alert one another to potential hazards like the approaching tail end of a traffic jam, accidents and icy conditions.

Figure 4.71 The Veniam software continuously searches for the best transmission technology. (Source: Bosch media).

Vehicles will also be able to take advantage of the traffic light 'green wave', because they will know when the next set of lights is going to turn green. The vehicles can then adjust their speed accordingly. This ensures the traffic flows more smoothly. Unfortunately, there is no globally harmonised standard for V2X communication currently available.

Key Fact

It is expected that the number of connected vehicles on the roads in Europe, the United States and China alone will exceed 470 million by 2025.

China primarily uses cellular-V2X technology (C-V2X), which is based on mobile communications; Europe and the United States are planning to additionally introduce transmission standards based on Wi-Fi (DSRC and ITS G5) alongside C-V2X. A mishmash of standards is therefore emerging internationally, which may lead to

vehicle communication issues. This is why the universal connectivity unit from Bosch is so important. Equipped vehicles will be able to communicate with one another as well as with their surroundings regardless of the vehicle make or the country in which they are used. This will make V2X communication even more secure and reliable.

The Veniam software is the connection enhancer for the connectivity unit from Bosch. As well as keeping an eye on which V2X communication technologies are currently available for use, the software also closely monitors the costs and data transmission latency of each alternative connection option, since not every technology is suitable in every situation.

For example, when it comes to alerting a driver to another vehicle that is about to pull out in front of them from a side street, every millisecond counts. This kind of critical information must be communicated in real time using highly reliable technology that is always ready for use, even if that means the resulting data transmission costs are

Figure 4.72 Connectivity unit. (Source: Bosch media).

Figure 4.73 V2V is one aspect of V2X connectivity

greater. Software updates from the cloud or a navigation system map update, on the other hand, can be put on hold in that sort of situation until a low-cost stationary Wi-Fi network becomes available. Large volumes of data can be transmitted via Wi-Fi in a short space of time, though a downside is that public or home Wi-Fi hotspots are not always available. Veniam's software is familiar with the pros and cons of each of the communication types and always establishes the optimal connection.

Key Fact

When alerting a driver to another vehicle that is about to pull out in front of them, every millisecond counts.

4.9.5 Motorcycle-to-car communication

Enabling motorcycles and cars to communicate with each other creates a digital shield for the motorcyclists. Vehicles within a radius of several hundred meters exchange information about vehicle type, speed, position and direction of travel, up to ten times a second.

Long before a motorcycle comes into view, this technology warns drivers and the sensors in their vehicles that a motorcycle is approaching. This allows them to drive better and more defensively. The public WLAN standard (ITS G5) is used as the basis for the exchange of data between motorcycles and cars. Transmission times of just a few milliseconds between transmitter and receiver mean that participating road users can generate and transmit important information relating to the traffic situation.

4.9.6 Cybersecurity

Introduction

Securing vehicles of the future is a cybersecurity challenge, to say the least. When something goes wrong with your home computer, 'crash' is only a metaphor.

A recent survey found that nearly 100% of today's cars include wireless technologies that could be insecure, and most manufacturers may not be able to easily determine if their vehicles have been hacked. Physical attacks via onboard diagnostic devices have shown it could be possible to manipulate some systems,

85

Figure 4.74 The vehicles know about each other before the drivers. (Source: Bosch media).

Figure 4.75 Cybersecurity is essential

steering for example, even while cars are moving.

Key Fact

Close to 100% of today's cars include wireless technologies that could be insecure.

Security, and specifically cybersecurity, is an increasingly urgent issue for the automotive industry. Systems are becoming more complex and the threat environment is also becoming more capable and sophisticated. This issue will only be made worse for ADVs because of V2X communication. A range of best practices exist ranging from management focus down to technical measures, which can help to control the risk.

The shift from independent, closed vehicle systems to one that is a connected environment is a massive change for the industry.

All vehicle systems much therefore have three mutually reinforcing properties:

- **Secure**: Prevention is better than a cure, and effective risk management begins by preventing system breaches in the first place.
- **Vigilant**: Hardware and software can degrade, and the nature and type of attacks can change. No level of security is perfect. Security must be therefore be monitored to ensure it is still secure or to see if it has been compromised.
- **Resilient**: When a breach occurs, there must be a system in place to limit the damage and re-establish normal operations. The system should also neutralise threats and prevent further spread.

The importance of securing individual sensors is critical in connected cars. Keep in mind these vehicles are a kind of internet connected data centre on wheels! A typical car can contain:

- About 70 computational systems running up to 100 million lines of code
- GPS devices that aid navigation and report on real-time traffic
- Diagnostic systems that check maintenance needs and send an alert in the event of an accident or breakdown.

As infrastructure evolves, cars will also be able to communicate with roadside devices such as traffic lights. Security must be part of the design and development, not bolted on at the end!

Key Fact

Security must be part of the design and development.

Key principles

As vehicles get smarter, cybersecurity in the automotive industry is becoming an increasing concern. Whether we're turning cars into Wi-Fi connected hotspots or equipping them with millions of lines of code to create fully autonomous vehicles, cars are more vulnerable than ever to hacking and data theft.

It's essential that all parties involved in the manufacturing supply chain, from designers and engineers, to retailers and senior level executives, are provided with a consistent set of guidelines that support this global industry. The UK Department for Transport (DfT), in conjunction with Centre for the Protection of National Infrastructure (CPNI), has created the following key principles for use throughout the automotive sector, the connected and autonomous vehicle (CAV) and intelligent transport systems (ITS) ecosystems and their supply chains:[9]

- Principle 1 – organisational security is owned, governed and promoted at board level
- Principle 2 – security risks are assessed and managed appropriately and proportionately, including those specific to the supply chain
- Principle 3 – organisations need product aftercare and incident response to ensure systems are secure over their lifetime
- Principle 4 – all organisations, including sub-contractors, suppliers and potential 3rd parties, work together to enhance the security of the system
- Principle 5 – systems are designed using a defence-in-depth approach
- Principle 6 – the security of all software is managed throughout its lifetime
- Principle 7 – the storage and transmission of data is secure and can be controlled
- Principle 8 – the system is designed to be resilient to attacks and respond appropriately when its defences or sensors fail

Figure 4.76 There are many aspects to cybersecurity

PAS 1885

To formalise the principles outlined previously, the specification PAS 1885:2018[10] has been written to help all parties involved in the vehicle lifecycle and ecosystem understand better how to improve and maintain vehicle security and the security of associated intelligent transport systems (ITS).

with the Centre for the Protection of National Infrastructure (CPNI). It sets out fundamental principles on how to provide and maintain cybersecurity in relation to reducing threat and harm to products, services and systems within increasingly connected and collaborative intelligent transport ecosystems.

> **Definition**
>
> PAS: A Publicly Available Specification, a standardisation document that closely resembles a formal standard in structure and format, but which has a different development model. The objective of a Publicly Available Specification is to speed up standardisation.

> **Definition**
>
> DfT: Department for Transport in the UK that works with agencies and partners to support the transport network.

This specification is based on the high-level set of guidelines (see previous section) developed by the UK Department for Transport (DfT) in conjunction

The concept of an automotive ecosystem encompasses:

▶ Vehicles
▶ Related infrastructure, including road-side and remote systems that provide services to the vehicles, their operators, occupants and cargo

▶ The human elements, including vehicle owners and/or operators, designers, manufacturers and service providers

This PAS applies to the security and functional safety aspects of the entire automotive development and use life cycle, including specification, design, implementation, integration, verification, validation, configuration, production, operation, servicing and decommissioning. A lifecycle approach is required to tackle all the risks that will arise from a constantly changing threat landscape, so as to protect vehicles and vehicle-related systems once they have been delivered to the market.

Hacking

The interconnectivity of current and future vehicles makes them potential targets for attack. Connectivity opens vehicle systems to the dark side of the internet, forcing automakers to quickly develop strategies to ensure that they don't join the litany of corporations hit by hack attacks. SAE Recommended Practice J3061, 'Cybersecurity Guidebook for Cyber-Physical Vehicle Systems', is the first document tailored for vehicle cybersecurity.

As more systems on vehicles are connected to the outside world by radio waves of some sort, or they scan the world outside of the car, then more opportunities are presented to hackers. Manufacturers are working very hard to reduce the chances of this happening and are helped in this process by what can be described as ethical hackers. There have been several interesting examples in the news recently; presented below are two examples that illustrate why this is an important area. In addition, there is currently a debate about the legality of hacking your own car as well as ways to stop other hackers.

> **Definition**
> Hacking: gaining unauthorised access to data in a system or computer.

Firstly, in 2015 Fiat Chrysler recalled 1.4 million vehicles in the USA because hackers had proved they could take control of an SUV over the internet and steer it into a ditch. Certain vehicle models manufactured from 2013 onwards required a software update to stop them from being controlled remotely. The two well-known experts in the area, Charlie Miller and Chris Valasek, performed the hack by breaking into the Jeep's UConnect system, which is designed to allow motorists to start their car and unlock the doors through an app.

The second example related to lidar used by autonomous vehicles. Jonathan Petit, Principal Scientist at the software security company Security Innovation, said he could take echoes of a fake car and put them at any location, and do the same with a pedestrian or a wall. Using such a system, with a cost of about US $50, that uses a kind of Laser pointer, attackers could trick a self-driving car into thinking something is directly ahead of it, causing it to slow down. Or by using numerous false signals, the car would not move at all.

Recently, vehicle owners and researchers in the US have been given the right to extract and examine car software without breaching copyright. The ruling follows the Volkswagen emissions scandal, in which software concealed higher than permitted emissions. In the UK, there are several laws that may limit the extent to which car owners can 'hack' the programming in their cars. These may be related to the copyright legislation concerning copying of code and evasion of technical protection measures, rights in confidential information and the Computer Misuse Act. Interestingly owners

89

don't normally own software, just the right to use it, but it is a grey area!

Connected vehicles bring many of the benefits gained with internet access, but they also bring security issues, including the threat of cyber-attacks. Design teams throughout the automotive supply chain therefore need to focus on a broad range of security technologies.

Key Fact

Connected vehicles bring the many benefits of internet access, but they also bring security issues, including the threat of cyber-attacks.

Security experts establish layers of protection so a threat that gets past one security barrier is stopped by another. Security should start with how the hardware for each ECU is designed and how the software is protected, and then extend to the communication between ECUs, to the design and segmentation of the vehicle network. The protections extend to those external inputs into the vehicle, Bluetooth, USB, OBD, the diagnostic port and the cellular network.

Definition

Hypervisor: A function which isolates operating systems and applications from the underlying computer hardware

Hypervisors are one technique for securing critical vehicle functions from errant or malicious software. They separate vehicle critical functions from user functions like infotainment, ensuring that problems that occur in radio head units don't spread.

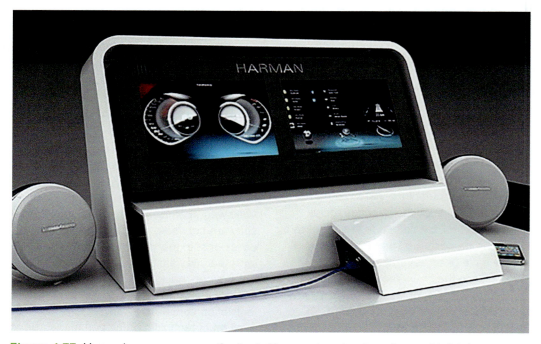

Figure 4.77 Hypervisors are among the tools Harman is using to safeguard infotainment systems. (Source: Harman).

A hypervisor can provide a secure isolation between an automotive domain and the user-facing operating system, minimising the risk of intrusions spreading to other systems.

Lidar and camera attacks

It has been shown in 2017 by a team at OnBoard Security (now part of Qualcomm Technologies, Inc.) that it was possible to trick sensors into thinking an object is in its path can cause a fake warning or trigger emergency braking.[11] One attack involved relaying the original signal sent from the vehicle lidar from a different position. This creates fake echoes and could make real objects appear closer or further than they really are. An extension of this attack was to actually create fake objects. The captured lidar signal was duplicated so objects could be re-created in any position.

Cameras can be blinded by shining bright light into the camera. This overexposes the image and hides the object from the ADV's system. Hitting the camera with bursts of light to confuse its controls, in some cases, resulted in the camera never recovering.

As ethical hackers, the OnBoard Security team proposed software and hardware countermeasures to improve sensors' resilience against these attacks. They have a goal to make CAVs as resistant to cyber-attacks as possible. The hope is that their research will identify potential attacks for automakers so they can make more robust systems and avoid potentially life-threatening situations for their customers.

4.10 Artificial intelligence (AI)

4.10.1 What is AI?

In computer science, artificial intelligence (AI) is intelligence demonstrated by machines, in contrast to the natural intelligence displayed by humans. Generally, the term 'artificial intelligence' is used to describe machines/computers that mimic things that humans associate with other human minds, such as learning and problem solving.

Some people prefer the term assistive intelligence (AI), because they suggest that machines are not intelligent, they just look like they are. We will leave this debate for another time!

There are four types of artificial intelligence as outlined in Table 4.2:

Figure 4.78 illustrates the routes taken towards different AI applications.

4.10.2 History of AI

The following is an interesting outline history of AI developments from Bosch.[12] Scientists have been working on artificial intelligence since the middle of the last century. Their goal was to develop machines that learn and think like humans. Here is an overview of the key learnings and technological milestones they have reached.

> **Key Fact**
>
> Scientists have been working on artificial intelligence since the middle of the last century.

1936: Turing machine

The British mathematician Alan Turing applied his theories to prove that a computing machine, known as a 'Turing machine', would be capable of executing cognitive processes, provided they could be broken down into multiple, individual steps and represented by an algorithm. In doing so, he laid the foundation for what we call artificial intelligence today.

Table 4.2 Types of artificial intelligence (AI)

Reactive machines	The most basic types of AI systems are purely reactive, and cannot form memories or use past experiences to inform current decisions. This type of intelligence involves the computer perceiving the world directly and acting on what it sees. It doesn't rely on an internal concept of the world. These machines behave exactly the same way every time they encounter the same situation. This is good when we want an AI system to be trustworthy, in an ADV for example. But it is not good if we want it to truly engage with, and respond to, the world.
Limited memory	This classification contains machines that can look into the past. Self-driving cars do some of this already. For example, they observe other cars' speed and direction. That can't be done in just one moment, but rather requires identifying specific objects and monitoring them over time. These observations are added to the self-driving cars' pre-programmed representations of the world, which also include lane markings, traffic lights and other important elements, like curves in the road. They're included when the car decides when to change lanes, to avoid cutting off another driver or being hit by a nearby car. But these simple pieces of information about the past are only transient. They aren't saved as part of the car's library of experience. Much research is happening in this area such that the ADVs AI system will remember and learn from previous experiences – a kind of AI 2.5!
Theory of mind (not yet complete)	Machines in this advanced class not only form representations about the world, but also about other agents or entities in the world. This is the understanding that people, creatures and objects in the world can have thoughts and emotions that affect their own behaviour. If AI systems are indeed ever to walk among us, they'll have to be able to understand that each of us has thoughts and feelings and expectations for how we'll be treated, and adjust their behaviour accordingly.
Self-awareness (does not yet exist)	The ultimate step in AI development is to make systems that can form representations about themselves. Consciousness even in humans is a difficult concept to define, and to build machines that have it even more of a challenge. It is also the point at which we should perhaps ask if we really want conscious machines?

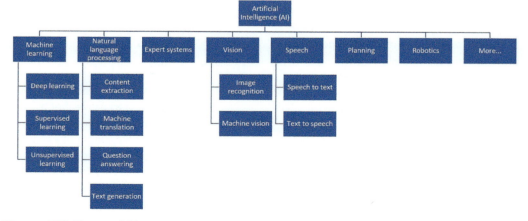

Figure 4.78 Types of AI

1956: The term AI was coined

In the summer of 1956, scientists gathered for a conference at Dartmouth College in New Hampshire. They believed that aspects of learning as well as other characteristics of human intelligence could be simulated by machines. The programmer John McCarthy proposed calling this 'artificial intelligence'. The world's first AI program, 'Logic Theorist', which managed to prove several dozen mathematical theorems and data, was also written during the conference.

1966: The first chatbot

The German-American computer scientist Joseph Weizenbaum of the Massachusetts Institute of Technology invented a computer program that communicated with humans. 'ELIZA' used scripts to simulate various conversation partners such as a psychotherapist. Weizenbaum was surprised at the simplicity of the means required for ELIZA to create the illusion of a human conversation partner.

1972: AI in the medical field

With 'MYCIN', artificial intelligence found its way into medical practices. The expert system developed by Ted Shortliffe at Stanford University was used for the treatment of illnesses. Expert systems are computer programs that bundle the knowledge for a specialist field using formulas, rules and a knowledge database. They are used for diagnosis and treatment support in medicine.

1986: NETtalk spoke

The computer was given a voice for the first time. Terrence J. Sejnowski and Charles Rosenberg taught their 'NETtalk' program to speak by inputting sample sentences and phoneme chains. NETtalk was able to read words and pronounce them correctly, and can apply what it has learned to words it didn't know. It was one of the early artificial neural networks, which are programs that are supplied with large datasets and are able to draw their own conclusions on this basis. Their structure and function are thereby similar to those of the human brain.

1997: Computer beat the world chess champion

The AI chess computer 'Deep Blue' from IBM defeated the incumbent chess world champion Garry Kasparov in a tournament. This was considered a historic success in an area previously dominated by humans. Critics, however, found fault with Deep Blue for winning merely by calculating all possible moves, rather than using cognitive intelligence.

2011: AI entered everyday life

Technology leaps in the hardware and software fields paved the way for artificial intelligence to enter everyday life. Powerful processors and graphics cards in computers, smartphones and tablets gave regular consumers access to AI programs. Digital assistants in particular enjoy great popularity: Apple's 'Siri' came to the market in 2011, Microsoft introduced the 'Cortana' software in 2014 and Amazon presented Amazon Echo with the voice service 'Alexa' in 2015.

2011: AI 'Watson' wins quiz show

The computer program 'Watson' competed in a US television quiz show in the form of an animated on-screen symbol and won against the human players. In doing so, Watson proved that it understood natural language and was able to answer difficult questions quickly.

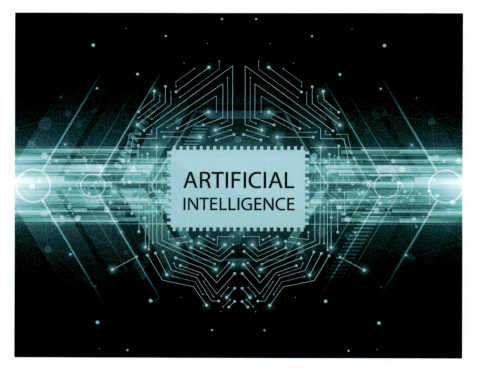

Figure 4.79 AI representation. (Source: www.vpnsrus.com).

2018: AI debated space travel and made a hairdressing appointment

These two examples demonstrated the capabilities of artificial intelligence. In June, 'Project Debater' from IBM debated complex topics with two master debaters, and performed remarkably well. A few weeks before, Google demonstrated at a conference how the AI program 'Duplex' phoned a hairdresser and conversationally made an appointment, without the lady on the other end of the line noticing that she was talking to a machine.

20xx: The near future is intelligent

Decades of research notwithstanding, artificial intelligence is comparatively still in its infancy. It needs to become more reliable and secure against manipulation before it can be used in sensitive areas, such as autonomous driving or medicine. Another goal is for AI systems to learn to explain their decisions so that humans can comprehend them and better research how AI thinks. Numerous scientists are working on these topics.

4.10.3 Top-down and bottom-up AI

There are advantages and disadvantages for both top-down and bottom-up AI systems. However, these tend to be complementary.

> **Key Fact**
> Artificial Intelligence (AI) can be described as top-down or bottom-up.

Top-down (symbolic) approach:

▶ Hierarchically organised (top down) architecture
▶ All the necessary knowledge is pre-programmed in the knowledge base
▶ Analysis involves creating, manipulating and linking symbols
▶ The program performs better at relatively high-level tasks such as language processing.

Bottom-up approach (neural networks for example)

▶ Models are built from simple components connected in a network
▶ Relatively simple abstract program consisting of learning cycles
▶ Program builds its own knowledge base and common sense assertions
▶ Intelligence emerges from the interactions of large numbers of simple processing units

▶ Built-in learning mechanism, results in adaptivity and flexibility
▶ Better able to model lower-level human functions, such as image recognition.

Two experts in the field warn that:

Rule based software can be a valuable part of a driverless car's toolkit for high-level control applications like route planning, and to manage low level activities, such as checking the status of the gas tank. However, rule based artificial intelligence has a tendency to break down in unstructured environments, leading some roboticists to refer to top down AI software as 'brittle'.

(Lipson and Kurman 2018)

Figure 4.80 shows a representation of methodologies and how or where they can be used.

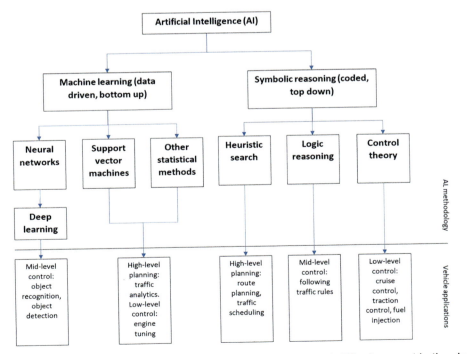

Figure 4.80 AI techniques used in driverless vehicles. The most difficult aspect is the deep learning required to recognise obstacles and to negotiate traffic. (Source: Bosch Media).

4.10.4 Deep learning

Deep learning is a class of machine learning algorithms that use multiple layers to progressively extract higher level features from a raw input. For example, in image processing, lower layers may identify edges, while higher layers may identify human-meaningful items such as a car, a person or an object (Figure 4.81).

> **Definition**
>
> Deep learning: A class of machine learning algorithms that use multiple layers to progressively extract higher level features from a raw input.

Multi-layer deeply-connected networks perform non-linear transformations on incoming data. It is often modelled on biological systems, such as deep neural networks (DNN) or convolutional neural networks (CNN). Systems are 'trained' by applying a large labelled training set in either a supervised or unsupervised optimisation process:

> **Definition**
>
> DNN: Deep neural networks

▶ Supervised machine learning is where the program is trained on a pre-defined set of examples, which then enable it to reach an accurate conclusion when given new data
▶ Unsupervised machine learning is where the program is given a large quantity of data and must find any patterns and relationships that it contains.

4.10.5 End to end machine learning

A machine learning end to end model learns all the features that can occur between the original inputs and the final outputs. For image recognition tasks, an end to end

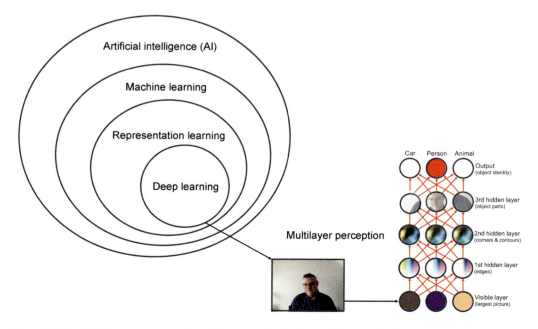

Figure 4.81 Multi-layer perception. (Source: Dr Mohamed Bergach, Kontron).

model is trained to recognise different items from an input image. The term end to end stresses that it handles the entire sequence of tasks, not part of the system. In this view, the additional steps, such as data collection or auxiliary processes, cannot be part of the model unless it can learn the intermediate processes with the given data.

Several ADV companies are working on end to end solutions using cameras as the only input. No lidars, no high-definition maps. One company in Cambridge, UK taught a car to drive itself in a simulated environment before unleashing it on the road equipped only with cameras and the normal satellite navigation system. The result was a self-driving car that could navigate complex new environments in a day.[13]

The company, Wayve.ai, is taking an end to end machine learning approach to building a self-driving car. Rather than create a platform and feed in the rules of the road, their system uses imitation and reinforcement learning coupled with cameras and sensors to control the vehicle and follow a route entered in the navigation system.

This model-based deep reinforcement learning system gives it the ability to learn to drive like a human in new environments based on data either given or learned from past experience. It's different from other autonomous platforms because the car creates all the rules itself and evolves based on safety driver interventions rather than relying on engineers to feed it new data.

Although the technology may perform well in real world scenarios, it's not clear it would have enough experience or how much it would take to appropriately handle edge cases, and the open road is not an acceptable testing ground to find out. Then there is the ability to figure out what went wrong in the event that, much like humans, it makes a mistake and causes an accident.

There is a view that these vehicles will be restricted to tightly geofenced areas, and it's not clear when the technology will be robust and trusted enough to be free to operate on all roads.

4.10.6 Object recognition simplified example

The processing power and complexity necessary to recognise objects in an image is massive. This section is a very much simplified explanation to show the process and what is involved.

Website

Visit www.automotive-technology.org for a working version of this and other simulation programs

To keep the process at a level we can work with, I have assumed that the images consist of grids of 5 × 5 pixels. Figure 4.82 shows four images used to train our AI. In reality a huge number of similar images (but much larger and more complex) would be used.

The images are processed to simplify them for example by looking at edges and darker areas. They are then stored in memory as a two-dimensional array, so the human, for example, is represented by:

$$Human01(5,5) = \begin{matrix} 0, & 0, & 1, & 0, & 0 \\ 0, & 0, & 1, & 0, & 0 \\ 0, & 1, & 1, & 1, & 0 \\ 0, & 1, & 1, & 1, & 0 \\ 0, & 1, & 0, & 1, & 0 \end{matrix}$$

Another human could be (in this case the same is pattern moved to the left):

$$Human01(5,5) = \begin{matrix} 0, & 1, & 0, & 0, & 0 \\ 1, & 1, & 1, & 0, & 0 \\ 0, & 1, & 0, & 0, & 0 \\ 1, & 1, & 1, & 0, & 0 \\ 1, & 0, & 1, & 0, & 0 \end{matrix}$$

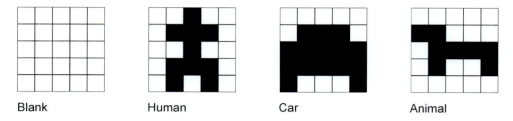

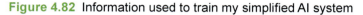

Blank   Human   Car   Animal

Figure 4.82 Information used to train my simplified AI system

The lines of code, and again I must stress this is much simplified, will be something like:

```
Call ProcessCurrentCameraImage(LiveImage(5, 5)) 'Stores the image from the camera
If LiveImage(5, 5) = Human01(5, 5) Or LiveImage(5, 5) = Human02(5, 5) Then
strObject = "Human"
ElseIf LiveImage(5, 5) = Animal01(5, 5) Or LiveImage(5, 5) = Animal02(5, 5) Then
strObject = "Animal"
ElseIf LiveImage(5, 5) = Car01(5, 5) Or LiveImage(5, 5) = Car02(5, 5) Then
strObject = "Car"
Else
strObject = "Unknown"
End If
```

There will be many more than two images used, and they can be iterated through in faster ways, but this illustrates the principle. Another thing that would be calculated (a change to the above program would be needed) is the percentage match. If the actual and stored images are identical, then the match will be 100%. If 1 of the 25 blocks in these examples was incorrect, then the match reliability would be 96% and so on.

It is possible therefore that the result we get from this updated program could be something like:

```
strObject = "Human" 96%
strObject = "Animal" 20%
strObject = "Car" 16%
```

The further processing has to deal with this imperfection. The more training that is done, then the more accurate the results will be.

Notes

1 The Highway Code states that flashing lights means 'I am here' or similar, but we all know the unwritten rule that it can also mean 'you go, I will slow down if needed'.
2 Source: LeddarTech, https://leddartech.com/
3 Source: https://autonomous-driving.org
4 Source: http://investors.tomtom.com/releasedetail.cfm?ReleaseID=1014506
5 Source: www.qualcomm.com/news/releases/2017/02/26/qualcomm-drive-data-platform-powers-tomtoms-plans-crowdsource-high
6 Assuming that high definition maps were available for the entire highway road network for the US, China, Germany and Japan (0.23M km), this represents less than 1% of the global road network (39.5M km).
7 Source: https://5g.co.uk/guides/what-is-5g/

8 Source: PwC. https://www.
  strategyand.pwc.com/media/file/2017-
  Strategyand-Digital-Auto-Report.
  pdf (September 2017); https://www.
  strategyand.pwc.com/media/file/Digital-
  Auto-Report-2018.pdf (September 2018)

9 Source: www.gov.uk/government/
  publications/principles-of-cyber-security-
  for-connected-and-automated-vehicles/the-
  key-principles-of-vehicle-cyber-security-for-
  connected-and-automated-vehicles

10 Source: https://shop.bsigroup.com/Produ
  ctDetail/?pid=000000000030365446

11 Source: http://blog.onboardsecurity.com/
  blog/author/jonathan-petit

12 Source: www.bosch.com/stories/history-
  of-artificial-intelligence/

13 Source: https://wayve.ai/

CHAPTER 5

Social and human issues

5.1 Who should die in a crash?

5.1.1 Classic trolley problem

The trolley (tram, train etc.) problem[1] is a thought experiment in ethics. The simple outline of the problem is as follows:

You see a runaway trolley moving toward five people lying on the tracks that are tied up so they can't move. You are standing next to a lever that controls a switch. If you pull the lever, the trolley will be redirected onto another track, and the five people on the main track will be saved. However, there is a single person lying tied up on the other track. You have two options:

▶ *Do nothing and allow the trolley to kill the five people on the main track*

▶ *Pull the lever, diverting the trolley onto the other track where it will kill one person.*

Which is the more ethical option? The dilemma can be adjusted, for example by saying that the single person on the track is your child or sister or similar.

It would appear that there is an obvious answer but when human emotions are considered then it is much more complex. The question that is asked in this type of scenario now in relation to ADVs is as follows: if forced to choose, who should a self-driving car kill or injure in an unavoidable crash?

Key Fact

It would appear that there is an obvious answer to the trolley problem but not when human emotions are involved.

Should the passengers in the vehicle be sacrificed to save pedestrians? Or should a pedestrian be killed to save a family of four in the vehicle? Of course, actual scenarios will be more complex, but it does illustrate how difficult it is to 'program' an automated driving vehicle (ADV) with ethics.

101

5 Social and human issues

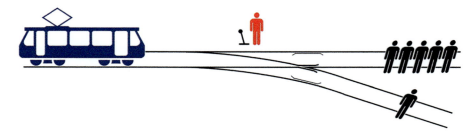

Figure 5.1 Who to kill? (Source: McGeddon, Wikimedia).

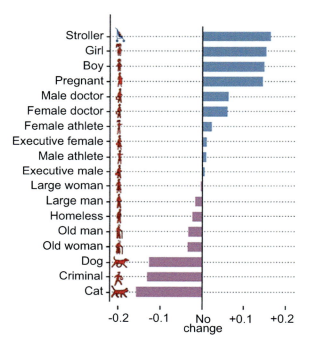

Figure 5.2 Should a self-driving car save passengers or pedestrians? (Source: MIT Media Lab).

Several car manufacturers have stated that the car will always look to save its occupants.

5.1.2 Self-driving car scenario

The Moral Machine[2] presented several variations of this dilemma involving a self-driving car. You can test your own reactions at their website.

People were presented with several scenarios. Should a self-driving car

sacrifice its passengers or swerve to hit (for example) a:

▶ baby
▶ male doctor
▶ criminal
▶ homeless person
▶ cat?

Four years after launching the experiment, the researchers published an analysis of the data (Figure 5.2). The results from 40 million decisions suggested people preferred to save humans rather than animals, spare as many lives as possible and tended to save young over older people.

There were also smaller trends of saving females over males, saving those of higher status over poorer people and saving pedestrians rather than passengers. The researchers acknowledge that their online game was not a controlled study and that it could not do justice to all of the complexity of autonomous vehicle dilemmas. However, they hope the Moral Machine will spark a conversation about the moral decisions self-driving vehicles will have to make.

Their view is that we need to have a global conversation to express our preferences to the companies that will design moral algorithms, and those who will regulate them. Germany has already introduced a law that states driverless cars must avoid injury or death at all costs. It also states that algorithms must never decide what to do based on the age, gender or health of the passengers or pedestrians.

If we are not vigilant, things may move in the direction of the carmakers rather than the public; we need to take care as Sperling considers: "We must steer oncoming innovations towards the public interest – if we don't, we risk creating a nightmare" (Sperling 2018).

5.2 Public reaction to CAVs

There are a number of hurdles to rolling out connected and autonomous vehicles (CAVs) in the UK and other countries.[3] Key challenges to adoption are:

Definition

CAV: Connected and autonomous vehicle

▶ Consumer behaviour
▶ Connectivity infrastructure
▶ Business model.

Increased acceptance of CAVs in society is likely to occur as their uptake gains momentum, first by influencers, and then by general consumers. Younger, more technology receptive people will make up a greater proportion of the driving population. Consumer confidence, as with everything else, will be greatly influenced by the media.

It is often perception rather than reality that sways public opinion. A self-driving car ran a red light during a trial in the US in 2016, which gained huge media attention. Imagine how many human drivers did the same during that time, with no media attention. As well as perceptions about safety, the consumer will also form opinions about cost. Caution must be taken that actual or perceived higher premiums in insurance do not hamper the uptake of CAVs.

This may also create an impression that the technology is less safe, which it is not. Imaging a world where we had never used petrol and then somebody invented a car that had 50 litres of it in a plastic tank at the back. It would be described as the most dangerous thing ever!

Communication from government and industry is needed to remove these barriers.

5.3 Insurance

The question has to be asked:

Who is responsible for an autonomous vehicle crash? It's not the driver! Drivers might not be terribly concerned with the distinctions between different levels of automation, but insurance companies most assuredly will be. Car insurance is expected to change radically in the era of self-driving cars.

(Sperling 2018)

The Association of British Insurers (ABI) has advised that driverless vehicles should have a sufficient level of security to guard against cyber-attacks before they are allowed to operate in fully autonomous mode. It also stressed that automated driving systems are able to detect and minimise the impact of cyber intrusions and data security breaches.

Definition

ABI: Association of British Insurers

Connected vehicle services bring lots of new features for the car and the driver, but could allow hackers to spread viruses or remotely access a vehicle's controls. Strong cybersecurity could become even more important than physical locks and immobilisers.

This point is one out of ten that insurers, and research body Thatcham Research, hope to see become a requirement for all driverless vehicles before they are allowed to operate in fully autonomous mode. Other recommendations include vehicle data being available after an accident, and vehicles being able to handle emergencies without driver intervention.

The ABI are major supporters of autonomous vehicles, because of the potential to dramatically improve road safety.

However, if people put their trust in a vehicle to get them safely from one place to another, building in appropriate cybersecurity is essential. In fact, it should be a compulsory requirement before any car is allowed to effectively drive itself.

5.4 Mobility as a service

Mobility-as-a-Service (MaaS) is potentially interesting shift away from vehicle ownership. It is thought we will buy a mobility solution rather than the actual vehicle. MaaS has many benefits that can, for example, improve road network efficiency, and the efficiency of energy usage.

This technology is a new way of thinking about transport. The key concept behind MaaS is to offer travellers mobility solutions based on their travel needs.

At first view these ideas seem ambitious to say the least; why don't we all use taxis like this now for example? However, compare how long your car sits on the drive or in the garage compared to driving on the road – it is a resource that is not being well used.

5.5 Global overview

5.5.1 United Kingdom
Introduction

A report in 2019 by the Society of Motor Manufacturers and Traders (SMMT)[4] offered a detailed assessment of connected and autonomous vehicle (CAV) development. The figures and core content in these

Figure 5.3 SMMT logo

sections are extracted from this report. Alternative views are added in some places. The full report is recommended reading and is available from the supplied weblink at the end of this chapter.

A key aspect was the deployment of CAVs in the UK. It covered three key aspects:

▶ Current market and technology trends, along with future roadmaps
▶ The UK's progress in, and propensity for, CAV deployment relative to other major countries
▶ The potential overall impact of CAV deployment on the UK's economy by 2030 and beyond.

A new deployment index, which benchmarks the UK and other major countries in terms of their progress toward CAV rollout was used. This comprehensive index is based on three macro parameters:

▶ Enabling Regulations
▶ Enabling Infrastructure
▶ Market Attractiveness.

Based on these three parameters, overall the UK comes out top, above rival nations,

including the US, Germany and Japan. This report also identified the economic benefit to the UK from the deployment of CAVs, which was estimated to be in the region of £62 billion per annum by 2030.

It concluded with an outlook to 2040, offering key recommendations to the UK government on how it can drive the opportunities presented by widespread CAV adoption. Figure 5.4 presents some of the key figures.

Over 18 million CAVs are expected to be added to the global motor parc by 2030. This will significantly change the way people commute.

Currently, from the five SAE levels, driver-in-the-loop assistance features, which are broadly level 2, are readily available on the market. These include features such as lane centring and adaptive cruise control.

Vehicles with higher levels of automation are set to become available during the 2020s. This is likely to start with driver-out-of-the-loop traffic jam and highway pilot features. These allow drivers to disengage safely from dynamic driving tasks (DDTs)

Figure 5.4 Benefits to the UK

such as manoeuvring in traffic jams and motorway driving.

Some early generation level 4 automation features may be introduced in the early 2020s. These could include highly automated highway pilot, automated valet parking and automated vehicles such as taxis operating within virtually defined or 'geofenced' zones in urban areas.

Level 5 automated vehicles have the capability to be fully self-driving, unconditionally, and with no operating domain or geographic restrictions. Level 5 is unlikely to be introduced before 2035. However, some commentators consider that the levels may remain mostly around 2 and then jump to level 5. This is because of the challenges in keeping the driver alert and ready to take control.

Key Fact

SMMT say level 5 is unlikely to be introduced before 2035; others suggest 2030.

The other view is that level 5 automated driving is likely to be reached gradually as more advanced driver assistance features come to market. This strategy, while incremental in its approach, is nonetheless expected to have a significant impact on the safety, convenience and cost aspects associated with current modes of transport.

Leading OEMs, including Audi, BMW, Ford, Jaguar Land Rover, Mercedes-Benz, Nissan, Tesla and Volvo, already provide level 2 driver assistance features in the UK. By 2030 it is expected that over 30% of all vehicles sold in the UK will be fitted with level 2 features.

Key Fact

30% of all vehicles sold in the UK will be fitted with level 2 features by 2030.

Both the roadmap for CAV deployment and real world applications for these technologies outlined in SAE J3016 (levels of automation) suggest that the first 'driver-out-of-the-loop' level 3 features such as traffic jam pilot and highway pilot could be deployed in the UK as early as 2021. We will see!

Support

The SMMT report makes four key recommendations to government:

- ▶ Amend road traffic laws to enable the deployment of level 3 AVs
- ▶ Ensure there is 4G coverage on the entire UK road network. This is essential for successful connected vehicles deployment. The current coverage of 4G on the UK's A and B road networks is limited to about 54%
- ▶ Encourage local authorities to work in collaboration with industry to implement consumer oriented urban mobility services, with safety as the central tenet of such initiatives
- ▶ Work towards an internationally harmonised set of regulations that define the testing, validation and type approval of AVs so industry can access global markets.

Economy

There is a potential annual economic benefit of £62 billion for the UK from CAV deployment by 2030. This is due to enhanced consumer productivity enabled by better in-car connectivity, improved travel efficiency and reduced mobility related expenses. For example, CAV deployment could result in commuters:

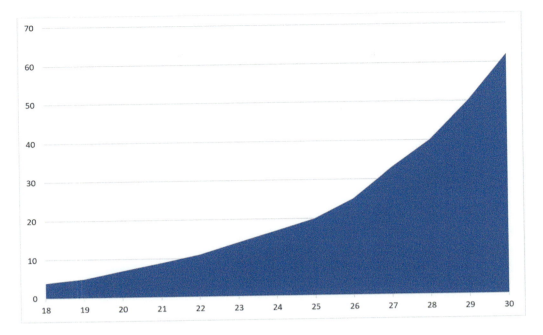

Figure 5.5 Economic benefits (£62 billion in 2030)

▶ Saving 42 hours per year travel time
▶ Average speeds per journey increased 20% due to reduced congestion and improved traffic flow.

It is suggested that there will be a wider impact of connectivity and automation on the economy due to new revenue streams. The automotive value chain is estimated to have a net positive impact on the economy. This is from mobility services, aftersales and vehicle insurance, but there will be new revenue sources for companies in this space. Licencing for hardware and software for CAVs is expected to be a major revenue generator because of new technology coming into the market.

Saving lives

Mobility convenience and reduction in overall costs are two of the key benefits of CAVs. However, the biggest impact on consumers is likely to be increased safety on the UK's road networks. Even basic driver assistance features such as automated emergency braking (AEB) and blind spot detection (BSD) are expected to significantly reduce the number of accidents.

However, by 2030, the overall benefits accrued from crash avoidance is estimated at more than £2 billion. This is due to a reduction of over 47,000 in serious collisions and a further 3,900 lives saved. Current projections suggest up to 56,000 crashes of all kinds can be eliminated by 2030 because of CAV related technologies. One in five miles travelled could be automated by 2030.

These figures are expected to be improved further when V2X applications, such as intersection collision warning and road hazard warning, add to the safety benefits of automation.

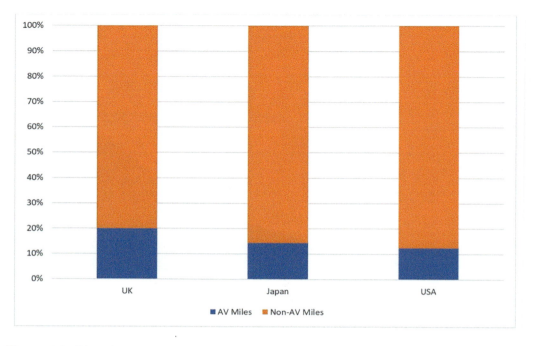

Figure 5.6 AV and non-AV miles by 2030

Definition

V2X: Vehicle to Everything

Summary

The SMMT report concludes by stating that the deployment of CAVs over the next decade is set to have a dramatic impact on both industry and the economy in the UK. The UK has the potential to emerge as a global centre of CAV development and deployment over the next decade and establish itself as one of the most attractive markets for CAV related investments. However, all of this will only be possible with active and sustained support from the government, especially in terms of investment in infrastructure and regulatory support.

5.5.2 European Union

At EU level, the focus on greener mobility has been joined by discussion on connected and automated mobility. In the Third Mobility Package, published on 17 May 2018, the European Commission included two key measures:

► Proposal for a revision of the General Safety Regulation (GSR) and Pedestrian Safety Regulation (PSR)
► Communication (non-legislative) on Connected and Automated Mobility (CCAM).

Since then, they have proposed a draft exemption to certification rules for automated vehicles and launched a roadmap for a future proposal on CCAM. The European Parliament are negotiating a non-legislative response to the European Commission's proposals.

The ongoing revision of the GSR/PSR includes a broad range of measures, the most significant of which include (level 2) active safety measures such as:

- Lane Keeping Assist
- Intelligent Speed Adaption
- Driver Distraction/Drowsiness Monitoring
- Autonomous Emergency Braking

The issue is unlikely to be agreed by the Council and the European Parliament in 2019, but it could potentially see some measures introduced in 2021. These could have a significant impact on road transport in the near future.

The European Commission (EC) have brought forward draft guidelines on providing vehicle certification exemptions, or type-approvals, for automated vehicles through a special compliance assessment procedure in respect of new technologies not yet covered by regulation. The EC also launched a public consultation on a roadmap at EU level relating to coordination between the Member States on the use of the 5G spectrum for testing connected cars, as well as sector-specific cybersecurity and data governance measures.

CAVs pose many difficulties for legislators, but there is a growing recognition of the advantages that the technology can bring, not least in terms of vehicle safety. Some key points are:

Key Fact

CAVs pose many difficulties for legislators, but there is a growing recognition of the advantages that the technology can bring, not least in terms of vehicle safety.

- EU legislators are turning towards autonomous technology as a long-term solution to help enhance vehicle and pedestrian safety.
- The European Union expects full self-driving capability by 2030.
- Mandated features on new cars could include lane-keeping technology, driver distraction sensors, external sensors, intelligent speed assistance and a 'black box' recorder which can be accessed to help determine causes of accidents.
- AV testing on public roads was legalised in several countries, including France, Germany, the Netherlands, Norway and the UK.

ACEA, the European Automobile Manufacturers' Association, has published a report in support of the EU's proposed revision of the General Safety Regulation. It noted in particular that active safety technologies, such as cameras and sensors, can reduce the number and severity of road accidents much more effectively than passive measures.

Definition

ACEA: European Automobile Manufacturers' Association

An analysis of road accident statistics was carried out by the Transport Research Laboratory (TRL) and Centre Européen d'Etudes de Sécurité et d'Analyse des Risques (CEESAR), and it offers guidance on the strengths and weaknesses of the safety measures proposed as part of the GSR revision.

The report found, for example, that systems to detect pedestrians and cyclists are 50% more effective in reducing fatalities and injuries than redesigning trucks to create 'Direct Vision' low-entry cabs. The report also stated that the benefits of reversing detection for trucks would be very limited in terms of accident mitigation. Extending front- and side-impact protection to vans and SUVs would also be limited, because the vehicles already have a high level of occupant protection.

Intelligent speed adaption (ISA) is when the speed of the vehicle is limited at all times and on all networks. The vehicle is informed

109

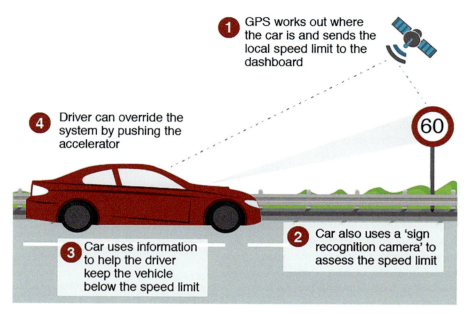

① GPS works out where the car is and sends the local speed limit to the dashboard

④ Driver can override the system by pushing the accelerator

60

③ Car uses information to help the driver keep the vehicle below the speed limit

② Car also uses a 'sign recognition camera' to assess the speed limit

Figure 5.7 Intelligent speed adaption. (Source: European Safety Council).

of the sign posted speed limit. It does not include dynamic speed limitation due, for example, to rain or fog. The system status is mandatory set to on. However, the driver can override the system by pushing harder on the accelerator.

> **Definition**
> ISA: Intelligent speed adaption

EU regulations and guidance continue to be developed. The UK is likely to be affected by these whether we are members of the EU or not.

5.5.3 United States of America

Some key points from recent changes (2018/19) in the USA are:

▶ California passes state level approval for driverless vehicle testing with no safety driver present.

▶ The US Department of Transportation issued guidance for automated driving pilot programmes.

▶ GM and Ford set up new automated driving divisions to accelerate AV deployment.

▶ Tesla rollout Level 2 and Level 2+ features with Level 3 and Level 4 features planned for 2020.

▶ Waymo, a spinoff from technology giant Google, starts first AV commercial business model in Arizona.

5.5.4 Japan and China

Some key points from recent changes (2018/19) in Asia are:

▶ Japan has considered policies related to liabilities, driving licenses and cybersecurity laws.

▶ China recently added 11 roads to the existing 33 in Beijing designated for ADV tests. ADVs are currently required to complete 5,000 km of daily driving in

designated closed test fields, before being allowed to apply for public road testing permits.
▶ China also granted permission to Audi, BMW and Daimler to test AVs in Beijing and Shanghai.

Notes

1 Philippa Ruth Foot, FBA, in 1967
2 Source: http://moralmachine.mit.edu
3 Source: www.smmt.co.uk
4 Source: www.smmt.co.uk/industry-topics/technology-innovation/connected-autonomous-vehicles/

CHAPTER 6

Case studies

6.1 Introduction

This case studies in this chapter are designed to give an overview of some assistive, connected, automated and autonomous vehicle technologies in use, or under development. Some are deliberately a few years old to illustrate how fast this world is are changing. Some of the materials are as supplied by the manufacturers, edited a bit for clarity and style.

6.2 Nvidia

Nvidia[1] is probably best known for computer graphics drivers, but they are a key player in the automated driving vehicle (ADV) market.

> **Key Fact**
> Nvidia are a key player in the automated driving vehicle market.

Conventional ADAS technology can detect some objects, do basic classification, alert the driver of hazardous road conditions and in some cases, slow or stop the vehicle.

This level of ADAS is ideal for applications like blind spot monitoring, lane change assistance and forward collision warnings.

NVIDIA DRIVE™ PX 2 AI car computers take driver assistance to the next level. They take advantage of deep learning and include a software development kit (SDK) for autonomous driving called DriveWorks. This SDK gives developers a powerful foundation for building applications that leverage computationally intensive algorithms for object detection, map localisation and path planning.

With NVIDIA self-driving car solutions, a vehicle's ADAS can discern a police car from a taxi; an ambulance from a delivery truck; or a parked car from one that is about to pull out into traffic. It can even extend this capability to identify everything from cyclists to absent-minded pedestrians.

NVIDIA DRIVE™ PX 2 is the open AI car computing platform that enables automakers and their tier one suppliers to accelerate production of automated and autonomous vehicles. It scales from a palm-sized, energy efficient module for AutoCruise capabilities, to a powerful AI

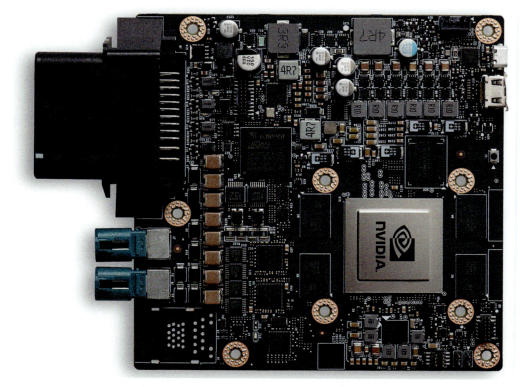

Figure 6.1 NVIDIA DRIVE PX 2 for AutoCruise. (Source: NVIDIA).

Figure 6.2 Pedestrian detection. (Source: NVIDIA).

supercomputer capable of fully autonomous driving.

The new single-processor configuration of DRIVE PX 2 for AutoCruise functions, which include highway automated driving and HD mapping, consumes just 10 W of power. It also enables vehicles to use deep neural networks to process data from multiple cameras and sensors.

DRIVE PX 2 can understand in real time what's happening around the vehicle, precisely locate itself on an HD map and plan a safe path forward. It is arguably the world's most advanced self-driving car platform, combining deep learning, sensor fusion and surround vision to change the driving experience.

The scalable architecture is available in a variety of configurations. These range from one passively cooled mobile processor operating at 10 W, to a multi-chip configuration with two mobile processors and two discrete GPUs delivering 24 trillion deep learning operations per second. Multiple DRIVE PX 2 platforms can be used in parallel to enable fully autonomous driving.

With a unified architecture, deep neural networks can be trained on a system in the data centre, and then deployed in the car.

Known as Drive PX, Nvidia's supercomputer for cars is powered by liquid-cooled Tegra X1 processors with two GPUs, which Nvidia claim to be as powerful as 150 MacBook Pros.

Nvidia's aim is to deliver a platform capable of a measure of artificial intelligence and situational awareness around the car, letting it navigate itself around hazards. The Drive PX 2 platform supports Nvidia DriveWorks, a suite of software tools, libraries and modules for developing and

Figure 6.3 Autonomous driving representation. (Source: NVIDIA).

Figure 6.4 NVIDIA's DGX SuperPOD. (Source: NVIDIA Media).

testing autonomous vehicles. The car needs to know exactly where it is, recognise the objects around it and continuously calculate the optimal path for a safe driving experience. The NVIDIA DriveWorks software development kit (SDK) gives developers a foundation upon which to build applications, using the computationally intensive algorithms for object detection, map localisation and path planning.

The first car maker to deploy Drive PX 2 was Volvo, which tested a fleet of 100 Volvo XC90s fitted with the system around Gothenburg, possibly the world's first public test of autonomous vehicles.

In June 2019 NVIDIA unveiled their 'Supercomputer to Develop Self-Driving Cars'.[2] It is currently the world's 22nd fastest supercomputer, a DGX SuperPOD, which provides AI infrastructure that meets the massive demands of the company's autonomous-vehicle deployment program.

Training the AI on self-driving cars is the ultimate computer-intensive challenge (see section 4.10 'Artificial intelligence (AI)'). A single data-collection vehicle generates 1 TB of data per hour. Multiply that by years of driving over an entire fleet, and you quickly get to petabytes of data. That data is used to train algorithms on the rules of the road,

and to find potential failures in the deep neural networks operating in the vehicle, which are then re-trained in a continuous loop.

> **Definition**
>
> Terabyte: A terabyte (TB) is a measure of computer storage capacity. It is approximately 2 to the 40th power, or 10 to the 12th power, or approximately a trillion bytes. A terabyte is more precisely defined as 1,024 gigabytes (GB), while a petabyte (PB) consists of 1,024 TB.

The new NVIDIA system is working around the clock, optimising autonomous driving software and retraining neural networks at a much faster turnaround time than previously possible. For example, the DGX SuperPOD hardware and software platform takes less than two minutes to carry out some tasks that in 2015 took 25 days to train on the then state-of-the-art system.

6.3 Bosch

6.3.1 (c2012)

High-performance assistance systems[3] already help drivers reach their destinations

Figure 6.5 Traffic jam assist. (Source: Bosch Media).

safely and more comfortably. Such systems control speed and the distance between vehicles. They also warn drivers of traffic jams and help them manoeuvre into even the tightest of parking spaces. Bosch, the global supplier of automotive technology and services, is set to expand its range of driver assistance systems in the years to come. In the future, these systems will take on a growing role in guiding vehicles through traffic jams. More specifically, they will brake, accelerate and steer completely autonomously. The traffic jam assistant will step in when the vehicle is moving at speeds between 0 and 50 kilometres per hour. This means that it will operate in most stop-and-go traffic situations. According to the German motor club ADAC, the total length of tailbacks in Germany alone amounted to 405,000 kilometres in 2011.

The first generation of the traffic jam assistant entered series production in 2014. In the following years, the feature was enhanced to cover ever-faster speeds and more complex driving situations. Eventually, the traffic jam assistant will serve as a highway pilot, making fully autonomous driving a reality.

Key Fact

The first generation of the traffic jam assistant entered series production in 2014.

Today, adaptive cruise control already tracks the vehicles ahead and adapts the distance and speed of the driver's own vehicle accordingly. Acting in combination

117

with the ESP® system and with the additional support of lane-detection cameras and electromechanical steering, this forms the technical basis for autonomous driving. High-performance software now calculates the appropriate driving instructions for a safer and less stressful driving. Automatic lane changing is the next functional step. It calls for two additional features. First, a rear-mounted radar sensor that also detects fast-approaching vehicles and, second, a dynamic navigation map. Such maps, which operate via a mobile network connection, can keep drivers informed of current roadwork sites and local speed restrictions. And although drivers remain responsible for driving, they can limit themselves to monitoring the actions of the driver assistance system.

Definition

ESP: Electronic stability program

As well as the ESP® and electrical steering, Bosch offers all the sensors required to detect the full range of traffic conditions relevant for drivers and their vehicles. Depending on the extent of onboard functions offered by a particular vehicle, front detection is carried out by a radar sensor combined with a mono camera, or by a stereo camera. With the LRR3, Bosch offers a high-performance long range radar sensor. With an aperture angle of up to 30 degrees, this sensor can detect objects at a distance of 250 meters. The new mid-range radar sensor, scheduled to go into series production in 2013, offers a range of 160 meters and an aperture angle of 45 degrees.

Its cost is significantly lower, since it is designed to meet the requirements of the mass market. In addition to the currently available multi-purpose video camera that is equipped with one sensor element, Bosch has developed a stereo video camera that

Figure 6.6 Bosch has been working on automated driving since 20110. (Source: Bosch Media).

detects objects in 3D with the help of two sensors. As a result, it is able to calculate exactly how far objects are from the vehicle, as well as in which direction they are moving. Both sensor configurations enable full predictive emergency braking.

Two adapted mid-range radar sensors assume the task of observing traffic behind the vehicle. These sensors have an aperture angle of 150 degrees and can detect objects up to 100 meters away. Finally, the parking assistant's ultrasound sensors provide support during close-range steering manoeuvres.

Level of automation continues to grow because assisting drivers in critical traffic situations can save lives. Drivers can also reach their destinations safely and with minimum stress using the Bosch traffic jam assistant. At speeds of up to 60 kph, the assistant brakes automatically in heavy traffic, accelerates and keeps the car in its lane.

Assistance systems are the cornerstone for automated driving, which will become established in a gradual process. Bosch already has its sights on highly automated driving, in which drivers no longer have to constantly monitor the vehicle.

With Bosch highway pilots, cars will be driving automatically on motorways by 2020, from entrance ramp to exit ramp. In the decade that follows, vehicles driving fully automated will be available, capable of handling any situations that arise.

Key Fact

ADAS are the cornerstone for automated driving, which will become established in a gradual process.

Automated driving affects every aspect of the car, such as powertrain, brakes and steering, and requires comprehensive systems expertise. It is based on sensors featuring radar, video and ultrasound technology. Powerful software and computers process the collected information and ensure that the automated vehicle can move through traffic in a way that is both safe and fuel efficient.

As vehicles gradually take over more and more driving tasks, safety-critical systems such as brakes and steering must satisfy special requirements. Should one of these components fail, a fallback is needed to ensure maximum availability. Bosch already has such a fallback for brakes: the iBooster, an electromechanical brake booster. Both iBooster and the ESP braking control system are designed to brake the car, independently of each other, without the driver having to intervene.

The Bosch iBooster meets an essential requirement for automated driving. The brake booster can build up brake pressure independently, three times faster than an ESP system. If the predictive brake system recognises a dangerous situation, the vehicle stops much faster. At the same time, the iBooster can also provide the gentle braking required by the adaptive cruise control (ACC), all the way down to a complete stop, and it is practically silent.

6.4 Google (Waymo)

6.4.1 (c2015)

The Google Self-Driving Car, commonly abbreviated as SDC, is a project by Google X that involves developing technology for autonomous electric cars.

In May 2014, Google presented a new concept for their driverless car[4] that had neither a steering wheel nor pedals, and unveiled a fully functioning prototype in December of that year that they planned to test on San Francisco Bay Area roads

119

Figure 6.7 Google car. (Source: Google/Waymo).

beginning in 2015. Google plans to make these cars available to the public in 2020.

Google's autonomous cars include about $150,000 in equipment, including a $70,000 lidar system. This is the range finder mounted on the top that uses a 64-beam laser. This laser allows the vehicle to generate a detailed 3D map of its environment. The car then takes these generated maps and combines them with high-resolution maps of the world, producing different types of data models that allow it to drive itself.

Heavy rain or snow produce safety concerns for all autonomous vehicle. Other issues are that the cars rely primarily on pre-programmed route data and as a result do not obey temporary traffic lights and, in some situations, revert to a slower 'extra cautious' mode in complex unmapped intersections.

The vehicle has difficulty identifying when objects, such as trash and light debris, are harmless, causing the vehicle to veer unnecessarily. Additionally, the lidar technology cannot spot some potholes or discern when humans, such as a police officer, are signalling the car to stop.

All developers of autonomous vehicles face these issues – Google aims to fix them by 2020. In June 2015, Google announced that their vehicles had driven over 1 million miles, and that in the process they had encountered 200,000 stop signs, 600,000 traffic lights and 180 million other vehicles.

6.5 Tesla Autopilot

6.5.1 (c2017)

Tesla Autopilot[5] is an increasingly capable suite of safety and convenience features that make personal transportation safer and more enjoyable. Since September 2014, Autopilot hardware has come standard in all Tesla vehicles, and Tesla has continued to refine and enhance the Autopilot system since its features were first enabled in cars

Figure 6.8 Tesla model 3. (Source: Tesla).

in October 2015 via over-the-air software updates. Data shows that, when used properly, drivers supported by Autopilot are safer than those operating without assistance. Eventually, full autonomy will enable a Tesla to be substantially safer than a human driver.

Key Fact

Since September 2014, Autopilot hardware has come standard in all Tesla vehicles.

In its current form, Autopilot is an advanced driver assistance system (ADAS) that classifies as a level 2 automated system. It is designed as a hands-on experience to give drivers more confidence behind the wheel, increase their safety on the road and make highway driving more enjoyable by reducing the driver's workload.

Autopilot's safety and convenience capabilities are designed to be additive to the driver's by augmenting their perception, improving their decision making and assisting in their control of the vehicle. Its user interface has been carefully designed to encourage proper use and to give drivers intuitive access to the information the car is using to inform its actions, via a detailed visual display on the instrument panel and clear audible cues. As Autopilot technology continues to be developed, more advanced functionality will be made available to Tesla owners over time nearing full self-driving capabilities; however, until truly driverless cars are developed and approved by regulators, the driver is responsible for and must remain in control of their car at all times.

Model S and Model X owners enjoy features like Autosteer, Auto Lane Change, Autopark and Summon, and Tesla is continuously innovating to keep customers at the forefront of technology through over-the-air software updates.

121

Key Fact

The driver is responsible for and must remain in control of their car at all times.

In October 2016, Tesla announced that all vehicles in production, as well as the forthcoming Model 3, will be built with an updated hardware suite, equipping each Tesla with the hardware needed for full self-driving capability at a safety level substantially greater than that of a human driver.

Eight surround cameras provide 360-degree visibility around the car at up to 250 meters of range. Twelve updated ultrasonic sensors complement this vision, allowing for detection of both hard and soft objects at nearly twice the distance of the prior system. A forward-facing radar with enhanced processing provides additional data about the world on a redundant wavelength that can see through heavy rain, fog, dust and even the car ahead. To make sense of all this data, a new on-board computer with over 40 times the computing power of the previous generation runs the new Tesla-developed neural net for vision, sonar and radar processing software. Together, this system provides a view of the world that a driver alone cannot access, seeing in every direction simultaneously and on wavelengths that go far beyond the human senses.

Key Fact

Eight surround cameras provide 360-degree visibility around the car at up to 250 meters of range.

Vehicles equipped with this hardware will continue to become more capable as new safety and convenience features are rolled out over time through over-the-air updates. These updates will significantly advance the driving experience and are available in two options:

▶ Enhanced Autopilot enables a Tesla to match speed to traffic conditions, keep

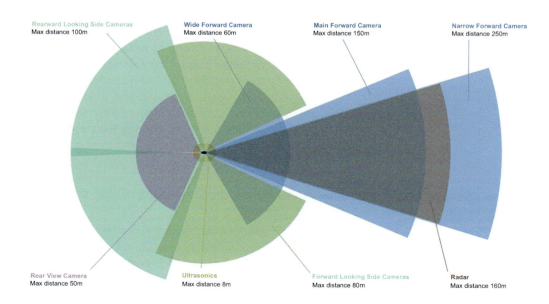

Figure 6.9 Tesla Autopilot sensor coverage. (Source: Tesla).

within a lane, automatically change lanes without requiring driver input, transition from one motorway to another, exit the motorway when your destination is near, self-park when near a parking spot and be summoned to and from your garage. Enhanced Autopilot should still be considered a driver's assistance feature with the driver responsible for remaining in control of the car at all times.

▶ Self-Driving, where permitted by regulatory approval, will ultimately take you from home to work and find a parking space for you on its own.

Before activating the features enabled by the new hardware, Tesla will further calibrate the new sensor suite using millions of miles of real-world driving to ensure significant improvements to safety and performance. While this is occurring, Teslas with new hardware will temporarily lack certain features currently available on Teslas with first-generation Autopilot hardware, including some standard safety features, such as automatic emergency braking, collision warning and Autopilot functionality, such as lane holding and active cruise control.

As these features are robustly validated, Tesla will enable them over the air, together with a rapidly expanding set of entirely new features. As always, their over-the-air software updates will keep customers at the forefront of technology and continue to make every Tesla, including those equipped with first-generation Autopilot and earlier cars, more capable over time.

Autopilot in vehicles built between September 2014 and October 2016 uses a combination of a camera and radar with enhanced processing, 12 ultrasonic sensors and navigation data to steer within a lane, changes lanes on prompt and adjusts speed in response to traffic. Autopilot features were first introduced to Tesla owners in October 2015 via a free over-the-air software update, and were found in several independent studies to be the most advanced level 2 driver assistance features available.

Figure 6.10 Autopilot in operation. (Source: Tesla).

Standard Autopilot features include:

▶ Autosteer and Traffic-Aware Cruise Control – Autosteer assists the driver on the road by steering within a lane. It relies on Traffic-Aware Cruise Control to maintain the car's speed in relation to surrounding traffic.

▶ Auto Lane Change – By engaging the turn signal when Autosteer is engaged, drivers can be assisted in transitioning to an adjacent lane on the right or left side of the car, when it is safe to do so.

▶ Autopark – When driving at low speeds on urban streets, a 'P' will appear on the instrument panel when a Tesla detects a parking spot. The Autopark guide will appear on the touchscreen along with the rear camera display, and, once activated, Autopark will begin to manoeuvre the vehicle into the parking space by controlling vehicle speed, gear changes and steering.

▶ Summon – With Summon, you can move Model S in and out of a parking space from outside the vehicle using the mobile app or the key.

6.6 Audi

6.6.1 (c2017)

With the Audi AI traffic jam pilot, the brand with the four rings presents the world's first system that enables SAE level 3 conditional automation. The car can take over the task of driving in a traffic jam or slow moving highway traffic up to 60 km/h (37.3 mph).

Key Fact

With traffic jam pilot engaged, drivers must still remain alert and capable of taking over the task of driving when the system prompts them to do so.

The driver activates the traffic jam pilot with the AI button on the centre console. On highways and multi-lane motorways with a physical barrier separating the two directions of traffic, the system takes over the driving task if the car is traveling at no more than 60 km/h (37.3 mph) in

Figure 6.11 Traffic jam pilot is active. (Source: Audi Media).

nose-to-tail traffic. The traffic jam pilot handles starting from a stop, accelerating, steering and braking in its lane. It can also handle demanding situations such as vehicles cutting in closely in front.

The control signals required by the system for conditional automated driving are obtained from the central driver assistance controller (zFAS) and from a redundant data fusion in the radar control unit.

If the traffic jam pilot is activated, drivers can take their foot off the accelerator and their hands off the steering wheel. Drivers must remain alert and capable of taking over the task of driving when the system prompts them to do so. They no longer have to continuously monitor the car and can focus on another activity supported by the on-board infotainment system, depending on the legal situation in the respective country.

The Audi virtual cockpit shows a stylised view of the car from the rear and blurred markings on the road that symbolise the motion and surroundings of the new A8. While traffic jam pilot is activated, a camera checks whether the driver is prepared to resume the task of steering if needed. It analyses the position and movement of the head and eyes in order to generate anonymised data.

If a driver's eyes remain closed for an extended period, for example, the system prompts the driver to resume the driving task. The prompt to take over is given in multiple stages. If the speed exceeds 60 km/h (37.3 mph) or the traffic begins to clear, the traffic jam pilot informs drivers they need resume driving themselves. If the driver ignores this prompt and the subsequent warnings, the A8 is braked until it stops completely in its lane.

Key Fact

While traffic jam pilot is activated, a camera checks whether the driver is prepared to resume the task of steering if needed.

Figure 6.12 Controls. (Source: Audi Media).

6 Case studies

Introduction of the Audi AI traffic jam pilot requires both clarity regarding the legal parameters for each individual country and specific adaptation and testing of the system. Moreover, varying worldwide homologation procedures and their deadlines must be observed. For these reasons, Audi will initiate series production of the traffic jam pilot in the new A8 incrementally, depending on the legal situation in the respective country.

Certain ambient conditions must be met for Audi AI traffic jam pilot to engage:

▶ The A8 is on a highway or a multi-lane road with barrier between oncoming lanes and a structure along the edge like guard rails.

▶ Slow-moving nose-to-tail traffic predominates in all neighbouring lanes.
▶ The vehicle's own speed must not exceed 60 km/h (37.3 mph).
▶ No traffic lights or pedestrians may be present within the relevant viewing range of the vehicle's sensors.

If these conditions are met, the driver sees a visual cue that the system is available: first, the Audi AI button on the console of the centre tunnel lights up in white. Next, a text message appears in the Audi virtual cockpit along with a pulsating white strip of light on its left and right edge. The AI icon in the digital instrument cluster also lights up in white.

Figure 6.13 Traffic jam pilot available. (Source: Audi Media).

Figure 6.14 Traffic jam pilot activated. (Source: Audi Media).

Once the driver has activated the traffic jam pilot by pressing the AI button, it lights up in green. The Audi virtual cockpit shows a stylised view of the car from the rear and blurred markings on the road. The vehicle's own speed appears digitally and in the form of a band on the bottom edge of the digital instrument cluster. Green edge lighting and the green AI icon symbolise the function.

While active, the Audi AI traffic jam pilot guides the new A8 within its own lane. The system manages starting from a stop, accelerating, steering and braking – the driver can relax. In this defined situation, the driver can take their foot off the accelerator and their hands off the steering wheel for longer periods and, in compliance with the applicable national regulations, can turn to other activities supported by the on-board infotainment system.

In Germany, for example, drivers have the option of watching TV programs and DVDs on the 10.1-inch display in the centre console and can use the Audi connect service to its fullest. They can turn their attention from the traffic and the car's steering to do things like answer their email, write text messages, tend their appointment calendar, read the news or plan for their vacation.

Thanks to its extensive sensor sets, the traffic jam pilot masters even demanding situations like vehicles cutting in closely in front. If the system detects an obstacle in front of the car, the A8 will avoid it if there is enough room within its lane to do so. Otherwise, it brakes the car to a standstill.

Key Fact

Thanks to its extensive sensor sets, the traffic jam pilot masters even demanding situations like vehicles cutting in closely in front.

The driving style of the traffic jam pilot is consistent and cooperative. During development of the system, special focus was placed on safety and comfort. Trials with numerous test subjects consistently led to the same result: people who use the traffic jam pilot quickly come to appreciate it. In traffic situations where driving is not much fun, it lets the driver relax and be chauffeured.

No value is placed on hectic lane changes. In fact, the system is not even designed for that: as soon as the driver sets the turn signal, the traffic jam pilot responds by prompting the driver to take over. The on-board monitor turns off the picture and the infotainment system lowers the volume. The driver indicates taking over the driving task by grasping the steering wheel, for example, which is detected by a capacity sensor. The steering torque sensor, gas pedal and brake pedal likewise register activity.

The A8 is equipped with driver readiness detection. While the traffic jam pilot is activated, it checks whether the driver is ready to retake the wheel. The system uses the camera installed in the top of the instrument panel. It analyses various criteria, including the position and movement of the head as well as monitoring the eyes. If the driver's eyes remain closed for a long period, for example, the system prompts the driver to prepare to resume driving. Activities not supported by on-board equipment, such as reading a newspaper, are generally not allowed. The camera's view of the driver's head becomes obstructed in this case, and the system will prompt the driver to take over.

The technical indicators generated by the image analysis software are anonymised, do not allow the face of the driver to be reconstructed and cannot be matched with any individual person. The data is processed locally within the car. The camera's images are not saved and there is no automatic transmission of data to AUDI AG or other third parties.

Figure 6.15 Traffic sign recognition. (Source: Audi Media).

When the traffic jam pilot prompts the driver to take over the driving task, the driver has about 10 seconds to respond, depending on the situation. In phase 1, a red light pulsates on the edge of the Audi virtual cockpit, the AI icon in the digital instrument cluster and the LED on the Audi AI button also light up in red and a subtle warning signal will sound. If the driver ignores this first prompt, this is followed by phase 2 – the "acute" warning. The warning signal becomes more distinct, the audio volume is lowered and the text "Traffic jam pilot: ending. Please resume full control of the vehicle!" appears in the Audi virtual cockpit. At the same time, the A8 slows down, gently at first and then with a jolt, and the driver feels the safety belt tighten slightly three times.

If the driver remains passive, due perhaps to a health issue, the final phase – emergency intervention – initiates. The warning signal becomes piercing, the safety belt is fully tightened. The A8 slows in its lane to a standstill and at the same time switches on the hazard lights. Once the sedan has come to a complete stop, the system activates the parking brake, shifts the Tiptronic to its P setting, unlocks the doors, switches on the interior light and then sends an emergency call via the mobile network if no response from the driver can be detected. This type of emergency stop in slow-moving traffic makes sense because it prevents the A8 from moving forward uncontrollably.

In trials with test subjects at Audi, most drivers responded during the first phase of the prompt to take over. The traffic jam pilot remains on standby until the driver switches it off with the Audi AI button. If the conditions are right for using it again, the system indicates its availability in the Audi virtual cockpit. The driver then merely needs to take their hands from the steering wheel to activate the traffic jam pilot.

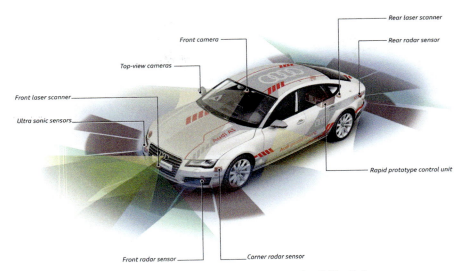

Figure 6.16 Audi A7 piloted driving concept. (Source: Audi Media).

While traffic jam pilot is activated, the speed of the Audi A8 is limited to 60 km/h (37.3 mph). If the traffic begins to clear and the vehicles ahead accelerate, the system remains active a few more seconds after a prompt to take over, until the driver has resumed driving.

In 2013, Audi was the first car manufacturer worldwide to obtain a testing license for the US states of California and Nevada. In January 2015, the Audi A7 piloted driving concept research vehicle drove 900 kilometres (559.2 mi) on the highway from San Francisco to Las Vegas. In May 2015, an automated Audi A7 drove in the dense urban traffic of Shanghai, China – a highly complex situation.

The traffic jam pilot handles starting from a stop, accelerating, steering and braking in its lane. Drivers no longer have to continuously monitor the car. When certain conditions are met, they can take their hands off the steering wheel for longer periods and can focus on other activities supported by the on-board infotainment system, depending on the legal situation in the respective country. As soon as the system reaches its limits, the car requires manual driver control again.

From a technical perspective, the traffic jam pilot is revolutionary. During piloted driving, a central driver assistance controller (zFAS) now continually computes an image of the surroundings by merging the sensor data. In addition to the radar sensors, a front camera and the ultrasonic sensors, Audi is the first car manufacturer also to use a laser scanner.

Key Fact

The new A8 is the world's first production automobile to have been developed specially for conditional automated driving at level 3 according to the applicable international standards.

6.7 Jaguar Land Rover

Drivers will be able to earn cryptocurrency and make payments on the move using

129

6 Case studies

innovative connected car services being tested by Jaguar Land Rover.

Using 'Smart Wallet' technology, owners earn credits by enabling their cars to automatically report useful road condition data such as traffic congestion or potholes to navigation providers or local authorities. Drivers could then redeem these for rewards such as coffee, or conveniently use them to automatically pay tolls, parking fees and for smart charging electric vehicles. 'Smart Wallet' removes the need for drivers to hunt for loose change or sign up to multiple accounts to pay for a variety of everyday services.

'Smart Wallet' uses the latest cryptocurrency technology and Jaguar Land Rover has partnered with the IOTA Foundation[6] to harness 'distributed ledger' technologies to make and receive these payments. Unlike other similar systems, due to its structure, it requires no transaction fee to operate and over time transactions will get faster across the entire network, forecast to include some 75 billion connected devices by 2025. Drivers could also top-up the 'Smart Wallet' using conventional payment methods.

Figure 6.17 JLR Connected cars. (Source: Jaguar Land Rover Media).

The advanced connected technology is being trialled at the new Jaguar Land Rover software engineering base in Shannon, Republic of Ireland, where engineers have already equipped several vehicles, including the Jaguar F-PACE and Range Rover Velar, with 'Smart Wallet' functionality.

The leading technology research forms part of Jaguar Land Rover's Destination Zero strategy which aims to achieve zero emissions, zero accidents and zero congestion. Part of reaching this target is developing a shared economy where the vehicle plays an integral role, as a data gatherer, in the smart city of the future. For example, the connected 'Smart Wallet' services will promote a reduction in congestion with the sharing of live traffic updates and offering alternative routes to drivers, reducing tailpipe emissions from idling in traffic.

Experts at the innovative Shannon R&D facility are developing new technologies to support electrification and self-driving features on future Jaguar and Land Rover vehicles. Supported by the Irish Development Agency, the Shannon team is also developing the next generation of electrical architecture as well as exploring advanced driver assistance systems features for future vehicles.

6.8 Toyota Guardian

6.8.1 (c2019)

Toyota Research Institute (TRI) showed a vivid re-enactment of a three-car crash on a California interstate, where no one was injured. A Toyota test vehicle observed the accident. Their test vehicle was travelling at freeway speed in manual mode with its autonomy mode disabled as it gathered data at the many tunnels and bridges in the San Francisco bay area. When the data was downloaded from the incident, the engineers asked a key question: could this crash have been mitigated, or avoided altogether by a future Toyota Guardian automated safety system?

Figure 6.18 Hidden vehicle coming from the side. (Source: Toyota Media).

Since 2016, TRI has been committed to a two-track development approach to automated driving. Its ongoing Chauffeur development focuses on full autonomy, where the human is essentially removed from the driving equation, either completely in all environments, or within a restricted operational design domain (ODD).

Toyota Guardian, on the other hand, is being developed to amplify human control of the vehicle, not replace it. With Toyota Guardian, the driver is meant to be in control of the car at all times, except in those cases where it anticipates or identifies a pending incident and employs a corrective response in coordination with driver input.

Key Fact

Toyota Guardian is being developed to amplify human control of the vehicle, not replace it.

One of TRI's most significant breakthroughs this year was the creation of *blended envelope control* where Guardian combines and coordinates the skills and strengths of the human and the machine. The system was inspired and informed by the way that modern fighter jets are flown, where you have a pilot that flies the stick, but actually they don't fly the plane directly. Instead, their intent is translated by the low-level flight control system, thousands of times a second to stabilise the aircraft and stay within a specific safety envelope.

This blended envelope control is much more difficult to create in a car than in a fighter jet. That is because the control envelope for a car is not only defined by vehicle dynamics, but also by the vehicle's perception and prediction ability of all things in its immediate environment.

The idea is that this control envelope is not a discrete on–off switch between the human and the autonomy system. It is a near-seamless blend of both, working as teammates to extract the best input from each.

Guardian is being developed as an automated safety system, capable of

Figure 6.19 Evasive action by Guardian. (Source: Toyota Media).

Figure 6.20 Potentially dangerous situation. (Source: Toyota Media).

Figure 6.21 Testing under a range of scenarios. (Source: Toyota Media).

operating with either a human driver, or an autonomous driving system, provided by Toyota, or any other company.

TRI stress the importance of not underestimating the difficulty of developing an autonomous Chauffeur system, both technologically and sociologically. Technically, the question is how do we train a machine about the social ballet required to navigate through an ever-changing environment, as well as, or better than, a human driver? Sociologically, public acceptance of the inevitable crashes, injuries and deaths that will occur due to fully autonomous Chauffeur systems may take considerable time.

133

In the meantime, TRI say they have a moral obligation to apply automated vehicle technology to save as many lives as possible as soon as possible.

That is why TRI's primary focus last year has been to concentrate most of its effort on making Toyota Guardian a smarter machine. For Guardian to learn and get smarter, it must be subjected to difficult and demanding driving scenarios, 'corner cases' that are simply too dangerous to perform on public roads. On closed courses, Guardian's intelligence and capabilities can be stretched and challenged. Through continuous refinement, Guardian learns how best to navigate and react to extremely dangerous scenarios, as they unfold.

This growing Guardian capability gives the three-car incident in California an interesting twist. Here was an accidental corner case on a public highway: a dangerous crash that unfolded right before Guardian's sensors and cameras. From that data, TRI developed an accurate simulation which was then translated into a learning tool for the car to figure out its options in a split-second. The scenario was then re-created on the test track, using real vehicles and a guided, soft-target, dummy-vehicle. In this instance, Guardian's best option was to quickly accelerate away from encroaching vehicles. Here is a case where Guardian might avoid, or mitigate a collision for itself, while potentially doing the same for other nearby vehicles.

6.9 FLIR

6.9.1 (2019)

FLIR[7] produces the only automotive-qualified thermal camera that is in cars today. More than 500,000 cars have been fitted with reliable night vision with pedestrian and animal detection. FLIR thermal cameras complement the ADAS and AV sensor suite. They provide the ability to reliably classify objects in the dark

Figure 6.22 Thermal cameras can see up to 4 times further than high beams illuminate. (Source: FLIR).

Figure 6.23 Working in difficult sunlight conditions. (Source: FLIR).

Figure 6.24 Reliably classify people and animals in cluttered environments. (Source: FLIR).

Figure 6.25 See further than high beams. (Source: FLIR).

and through obscurants, including smoke, sun glare and most fog – day and night. Because they see heat, they have a unique ability to reliably classify people and animals better than other ADAS and AV sensor technologies. Including thermal cameras in the suite of sensors increases the situational awareness, reliability and safety.

Figure 6.26 PathFindIR Camera. (Source: FLIR).

PathFindIR II is a powerful thermal night vision system that helps to see road hazards in total darkness, and it will alert drivers to nearby vehicles, people and animals. Headlights typically illuminate about 400 m straight ahead, but PathFindIR II detects heat without a need for light, allowing the car to see up to four times further. It provides visibility through dust, smoke or fog, and better capability to avoid an accident.

Figure 6.27 Camera. (Source: First Sensor AG).

systems. The company develops and manufactures standardised and customised sensor solutions for applications in the industrial, medical and mobility growth markets.

They produce a range of sensors but of interest here are the cameras and ADAS. Systems for ADAS include:

6.10 First sensor AG

6.10.1 (2019)

First Sensor AG is one of the world's leading suppliers in the field of sensor

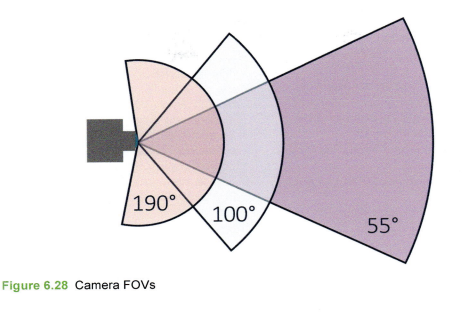

Figure 6.28 Camera FOVs

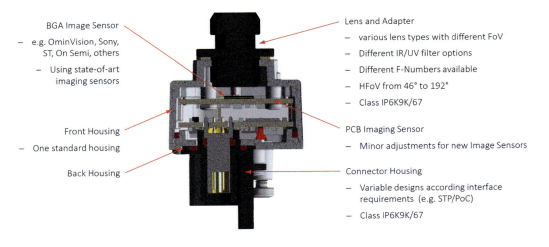

BGA Image Sensor
- e.g. OminVision, Sony, ST, On Semi, others
  - Using state-of-art imaging sensors

Front Housing
- One standard housing

Back Housing

Lens and Adapter
- various lens types with different FoV
- Different IR/UV filter options
- Different F-Numbers available
- HFoV from 46° to 192°
- Class IP6K9K/67

PCB Imaging Sensor
- Minor adjustments for new Image Sensors

Connector Housing
- Variable designs according interface requirements (e.g. STP/PoC)
- Class IP6K9K/67

Figure 6.29 Camera details. (Source: First Sensor AG).

- Embedded ECU as platform for Sensor Fusion
- Inhouse software development and testing
- Full system capability for commercial, special vehicles and mobile machines
- Integration of lidar, radar, camera, ultrasonic sensors

Specifications of one of the latest cameras are typically as follows:

- Increased robustness and long durability
- Automotive qualification tests for temperature, electro-magnetic resistance, vibration, etc.
- IATF 16949 certification

137

- ▶ High resolution and sensitivity (low-light performance down to 0.05 lux)
- ▶ High dynamic range (HDR) – up to 132 dB
- ▶ Precise lens/image sensor alignment process
- ▶ Good image quality in all areas of the image, minimum de-centre
- ▶ High-resolution cameras with less than 2 W power demand
- ▶ Different fields of view (FOV: 55°/100°/190°)
- ▶ Various interfaces (e.g., APIX, LVDS, Ethernet. . .)

Notes

1 Source and for more information, visit: www.nvidia.com/en-us/self-driving-cars/drive-platform/
2 Source: https://blogs.nvidia.com/blog/2019/06/17/dgx-superpod-top500-autonomous-vehicles/
3 Source: www.bosch-presse.de/presseforum/detail/en-US&txtID=6071
4 Source: www.google.com/selfdrivingcar
5 Source: www.tesla.com
6 Source: www.iota.org/get-started/what-is-iota
7 Source: www.flir.co.uk/oem/adas/

CHAPTER 7

Website

7.1 Introduction

The online Automotive Technology Academy has been created by the author (Tom Denton), who has over 40 years of relevant automotive experience, and over 20 published textbooks that are used by students and technicians worldwide.

The aims of the online academy are to:

▶ Improve automotive technology **skills** and **knowledge**
▶ Provide **free** access to study resources to support the textbooks
▶ Create a worldwide **community** of automotive learners
▶ Freely **share** automotive related information and ideas
▶ **Reach** out to learners who are not able to attend school or college
▶ Improve automotive training **standards** and **quality**
▶ Provide online access to **certification** for a range of automotive subjects

To access the academy visit: www.automotive-technology.org and create an account for yourself. To access the free courses, which work in conjunction with the textbooks, you will need to enter an enrolment key. This will be described something like this:

Website

www.automotive-technology.org

'The third word on the last line of page ## of the associated textbook'. You will therefore need to own the book!

All you need to do is enter the word in a box and you will have full unrestricted access to the course and its associated resources.

7.1.1 Resources

The following is a list of some of the resources that will be available to you:

▶ Images
▶ Videos
▶ Activities
▶ 3D models
▶ Hyperlinks
▶ Assignments
▶ Quizzes

Figure 7.1 Audi RS7 Concept car. (Source: Audi Media).

▶ Forums
▶ Chat features
▶ Social media
▶ Interactive features, games and much more.

A progress bar is used in the courses, so you can see at a glance how you are getting on, if you are working to cover all the content. Alternatively, you can just dip in and out to find what you need.

More formal assessments will also be available for those who are not able to attend traditional training centres. It will be possible to obtain certification relating to theory and practical work. A charge will apply to this aspect, but most other resources are free.

Updates and interesting new articles will also be available, so what are you waiting for? Come and visit and join in!

References

Dalton, J. (2019). "Thatcham Research warns over 'autonomous' vehicle marketing". Source: www.driverlessguru.com/blog/thatcham-research-warns-over-autonomous-vehicle-marketing.

Denton, T. (1995). *Automobile Electrical and Electronic Systems*. London: Edward Arnold.

Lipson, H. a. and M. a. Kurman (2018). *Driverless: Intelligent Cars and the Road Ahead*. Cambridge, MA: Massachusetts Institute of Technology Press.

McGrath, M. E. (2018). *Autonomous Vehicles*. Amazon.

Herrmann, A., Brenner, W. and Stadler, R. (2018). "Prelims". *Autonomous Driving*. Emerald Publishing Limited, pp. i–xiv. https://doi.org/10.1108/978-1-78714-833-820181046

SAE. (2018). "Taxonomy and definitions for terms related to driving automation systems for on-road motor vehicles". *SAE*. https://www.sae.org/standards/content/j3016_201806

Sperling, D. e. (2018). *Three Revolutions: Steering Automated, Shared, and Electric Vehicles to a Better Future*. Washington, DC: Island Press.

Index

Index